Life Science

_____ Grade 7 _____

Written by Tracy Bellaire

The experiments in this book fall under eight topics that relate to two aspects of life science: **Interactions Within Ecosystems in the Environment; and Plants for Food & Fibre.** In each section you will find teacher notes designed to provide you guidance with the learning intention, the success criteria, materials needed, a lesson outline, as well as provide some insight on what results to expect when the experiments are conducted. Suggestions for differentiation are also included so that all students can be successful in the learning environment.

Tracy Bellaire is an experienced teacher who continues to be involved in various levels of education in her role as Differentiated Learning Resource Teacher in an elementary school in Ontario. She enjoys creating educational materials for all types of learners, and providing tools for teachers to further develop their skill set in the classroom. She hopes that these lessons help all to discover their love of science!

Published in Canada by:
On The Mark Press
Belleville, ON
www.onthemarkpress.com

Funded by the
Government
of Canada

Canadä

OTM2166   ISBN: 9781487710361
© On The Mark Press

At A Glance

Learning Intentions

Knowledge and Understanding Content	Ecosystems and Biomes	Ecosystems at Work!	Roles in the Natural World	Food Chains and Webs	Ecological Succession	Human Actions and Technology	Plants – At the Root of It!	Soils and Plant Growth
-describe the common characteristics of an ecosystem and identify the major biomes of the world; research a biome	•							
-recognize biotic and abiotic components and describe how they interact within ecosystems; conduct a field study		•						
-recognize roles in food chains, explain the balance plants and animals provide in the ecosystems; create a composter			•					
-identify, explain, and build food chains and webs consisting of different living things; describe their interactions				•				
-describe primary and secondary succession; create an ecosystem that is able to sustain itself and grow					•			
-recognize the impact human actions and technologies have on ecosystems in the environment; determine an action plan						•		
-recognize the importance of plants in our daily lives and identify links between technologies, products, and impacts							•	
-explore the components in soils, their compaction and moisture content; recognize the main parts of a plant; demonstrate seed germination and healthy plant growth								•
Thinking Skills and Investigation Process								
-make predictions, formulate questions, and plan an investigation		•	•		•			•
-gather and record observations and findings using drawings, tables, written descriptions	•	•	•	•	•	•	•	•
-recognize and apply safety procedures in the classroom	•	•	•	•	•	•	•	•
Communication								
-communicate the procedure and conclusions of investigations using demonstrations, drawings, and oral or written descriptions, with use of science and technology vocabulary	•	•	•	•	•	•	•	•
Application of Knowledge and Skills to Society and the Environment								
-assess the impact of human activities and technology on the environment and ecosystems						•	•	
-analyze the costs and benefits of certain strategies for protecting the environment and achieving a level of sustainable development						•	•	
-analyze the technologies used to enhance soils, seed germination, and healthy plant production in the farming industry								•

OTM2166   ISBN: 9781487710361
© On The Mark Press

TABLE OF CONTENTS

OTM2166   ISBN: 9781487710361
© On The Mark Press

Teacher Assessment Rubric

Student's Name: _____   Date: _____

Success Criteria	Level 1	Level 2	Level 3	Level 4
Knowledge and Understanding Content				
Demonstrate an understanding of the concepts, ideas, terminology definitions, procedures and the safe use of equipment and materials	Demonstrates limited knowledge and understanding of the content	Demonstrates some knowledge and understanding of the content	Demonstrates considerable knowledge and understanding of the content	Demonstrates thorough knowledge and understanding of the content
Thinking Skills and Investigation Process				
Develop hypothesis, formulate questions, select strategies, plan an investigation	Uses planning and critical thinking skills with limited effectiveness	Uses planning and critical thinking skills with some effectiveness	Uses planning and critical thinking skills with considerable effectiveness	Uses planning and critical thinking skills with a high degree of effectiveness
Gather and record data, and make observations, using safety equipment	Uses investigative processing skills with limited effectiveness	Uses investigative processing skills with some effectiveness	Uses investigative processing skills with considerable effectiveness	Uses investigative processing skills with a high degree of effectiveness
Communication				
Organize and communicate ideas and information in oral, visual, and/or written forms	Organizes and communicates ideas and information with limited effectiveness	Organizes and communicates ideas and information with some effectiveness	Organizes and communicates ideas and information with considerable effectiveness	Organizes and communicates ideas and information with a high degree of effectiveness
Use science and technology vocabulary in the communication of ideas and information	Uses vocabulary and terminology with limited effectiveness	Uses vocabulary and terminology with some effectiveness	Uses vocabulary and terminology with considerable effectiveness	Uses vocabulary and terminology with a high degree of effectiveness
Application of Knowledge and Skills to Society and Environment				
Apply knowledge and skills to make connections between science and technology to society and the environment	Makes connections with limited effectiveness	Makes connections with some effectiveness	Makes connections with considerable effectiveness	Makes connections with a high degree of effectiveness
Propose action plans to address problems relating to science and technology, society, and environment	Proposes action plans with limited effectiveness	Proposes action plans with some effectiveness	Proposes action plans with considerable effectiveness	Proposes action plans with a high degree of effectiveness

OTM2166   ISBN: 9781487710361
© On The Mark Press

Student Self Assessment Rubric

Name: _____ Date: _____

Put a check mark ✔ in the box that best describes you:

	Always	Frequently	Sometimes	Seldom
I listened to instructions.				
I was focused and stayed on task.				
I worked safely.				
My answers show thought, planning, and good effort.				
I reported the results of my experiment.				
I discussed the results of my experiment.				
I used science and technology vocabulary in my communication.				
I connected the material to my own life and the real world.				
I know what I need to improve.				

1. I liked _____

_____

2. I learned _____

_____

3. I want to learn more about _____

_____

OTM2166   ISBN: 9781487710361
© On The Mark Press

INTRODUCTION

The activities in this book have two intentions: to teach concepts related to life science and to provide students the opportunity to apply necessary skills needed for mastery of science and technology curriculum objectives.

Throughout the experiments, the scientific method is used. The scientific method is an investigative process which follows five steps to guide students to discover if evidence supports a hypothesis.

1. **Consider a question to investigate.**
   For each experiment, a question is provided for students to consider. For example, "What effect could the introduction of a new plant or animal species have on an ecosystem?"

2. **Predict what you think will happen.**
   A hypothesis is an educated guess about the answer to the question being investigated. For example, "I believe that introducing a new plant or animal to an established ecosystem could upset the balance, and endanger the existence of another species common to the particular ecosystem". A group discussion is ideal at this point.

3. **Create a plan or procedure to investigate the hypothesis.**
   The plan will include a list of materials and a list of steps to follow. It forms the "experiment".

4. **Record all the observations of the investigation.**
   Results may be recorded in written, table, or picture form.

5. **Draw a conclusion.**
   Do the results support the hypothesis? Encourage students to share their conclusions with their classmates, or in a large group discussion format.

The experiments in this book fall under eight topics that relate to two aspects of life science: **Interactions within Ecosystems in the Environment, and Plants for Food and Fibre.** In each section, you will find teacher notes designed to provide you guidance with the learning intention, the success criteria, materials needed, a lesson outline, as well as provide some insight on what results to expect when the experiments are conducted. Suggestions for differentiation are also included so that all students can be successful in the learning environment.

ASSESSMENT AND EVALUATION:

Students can complete the Student Self-Assessment Rubric in order to determine their own strengths and areas for improvement. Assessment can be determined by observation of student participation in the investigation process. The classroom teacher can refer to the Teacher Assessment Rubric and complete it for each student to determine if the success criteria outlined in the lesson plan has been achieved. Determining an overall level of success for evaluation purposes can be done by viewing each student's rubric to see what level of achievement predominantly appears throughout the rubric.

OTM2166   ISBN: 9781487710361
© On The Mark Press

ECOSYSTEMS AND BIOMES

LEARNING INTENTION:

Students will learn about the characteristics of an ecosystem and the major biomes of the world.

SUCCESS CRITERIA:

- define the meaning of ecosystem and identify specific examples of ecosystems
- recognize an ecosystem, determine what makes it an ecosystem
- identify some common characteristics of ecosystems
- identify and describe the major biomes of the Earth
- research a biome to determine its location(s), its climate, vegetation, animals home to it
- research and report three interesting facts about a chosen biome

MATERIALS NEEDED:

- a copy of "Defining an Ecosystem" worksheet 1, 2, and 3 for each student
- a copy of "What Makes It An Ecosystem?" worksheet 4 for each student
- a copy of "Biomes of the Earth" worksheet 5, 6, and 7 for each student
- a copy of "Researching a Biome" worksheet 8, 9, and 10 for each student
- access to the internet or local library
- dictionaries
- clipboards, pencils, pencil crayons, markers, chart paper
- access to a printer (*optional*)
- modeling clay, construction paper, glue, scissors, shoe boxes for dioramas (*optional*)

PROCEDURE:

*This lesson can be done as one long lesson, or be done in four or five shorter lessons.

1. Divide students into pairs and give each pair worksheet 1. They will engage in a 'Think-Pair-Share' activity to discuss the definition of an ecosystem. Then come back together as a large group to discuss and record their ideas on chart paper.

2. Give students worksheets 2 and 3 to complete. They will need to work with a partner again to complete a section on worksheet 2. An option is to come back as a large group to orally share ideas of examples of specific ecosystems.

3. Give students worksheet 4, and a pencil and clipboard. With access to the internet, students will search for an example of an ecosystem (diagram or picture). They can either illustrate it on the worksheet space provided, or print out a copy and paste it in place. They will share and compare their drawings with a classmate, then discuss the common characteristics of an ecosystem. Students should realize that:

- an ecosystem contains living and non-living things
- it is ever changing, growing, reducing, without limits
- there are different types of ecosystems in different areas, each with a purpose
- it is an integration of living and non-living things, dependent upon each other
- it is a functional unit, working to meet needs

4. Using worksheet 5, 6, and 7, do a shared reading with the students. Along with the content, discussion of some vocabulary would be beneficial for students so they can understand the passage.

Some interesting vocabulary words to focus on are:

climate	semiarid desert
equator	greenhouse gas
characterized	coastal desert
tropical	coniferous forest
tundra	hibernation
nourishing	deciduous forest
humid	vegetation
savanna	temperate grasslands

5. To enhance students' learning about the Earth's biomes, have them watch either 'Biomes Our Earth's Major Life Zones (fixed)' or 'Biomes of the World for Children: Oceans, Mountains, Grassland, Rainforest, Desert'. Both episodes can be accessed at www.youtube.com.

6. Give students worksheets 8, 9, and 10. They will choose a biome that they would like to learn more about. They will describe its location(s), indicate it on a world map, describe the climate, identify two types of vegetation and two animal species that are native to the biome. They will research three interesting facts about this biome. Students may need to access the internet or local library to obtain information. A follow up option is to have students share their fast facts with a classmate or small group.

DIFFERENTIATION:

Slower learners may benefit by:

• working in a small group with teacher direction to complete worksheets 1 and 2. Students' responses about the meaning of an ecosystem and specific examples could be recorded on chart paper. An option is to share these responses with the large group for added discussion.

• a reduction of expectation by either eliminating worksheet 10, or requiring theses students to research only one interesting fact

For enrichment, faster learners could design a representation of the biome they chose on worksheet 8. This could be done by making a diorama. Dioramas could be displayed in the classroom, or somewhere in the school for others to observe. Students could also provide a written component to accompany their dioramas, which would detail the location(s), climate, and interesting facts about it.

OTM2166   ISBN: 9781487710361
© On The Mark Press

Name:

Defining an Ecosystem

Think **Pair** **Share**

With your partner, do some thinking and sharing of ideas about the question below.

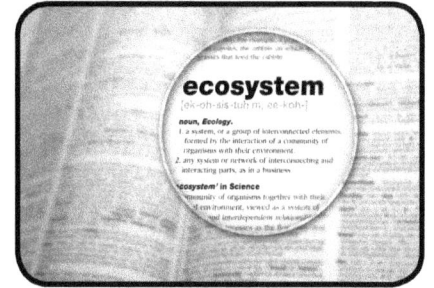

"What is an ecosystem?"

Record Your Thinking!

OTM2166   ISBN: 9781487710361
© On The Mark Press

Use a dictionary to look up the meaning of the word 'ecosystem'. Record the definition below.

Ecosystem: _____

_____

_____

_____

With your partner, continue to brainstorm some *specific examples* of ecosystems. Record your ideas in the box below.

OTM2166 ISBN: 9781487710361
© On The Mark Press

Name:

Is this an ecosystem? Explain your thinking.

Is this an ecosystem? Explain your thinking.

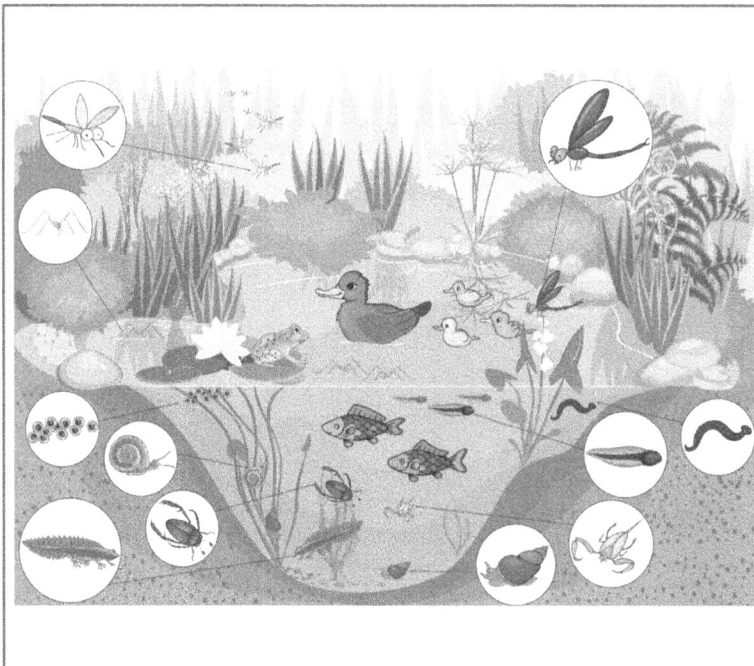

OTM2166   ISBN: 9781487710361
© On The Mark Press

What Makes it an Ecosystem?

Access the internet to look for examples of ecosystems. Once you find one, illustrate it in the box below, or print a picture of it and paste it in the box below.

Share and compare your image with a classmate's findings.

What are some common characteristics of ecosystems that you notice?

_____

_____

_____

_____

_____

OTM2166   ISBN: 9781487710361
© On The Mark Press

Biomes of the Earth

A community of plants and animals existing in a region of the Earth, within a certain climate type is called a **biome**. Biomes can be compared to very large ecosystems, in which, are smaller ecosystems. There are many different biomes on our Earth. Some of the major ones are deserts, forests, grasslands, tundra, and aquatic biomes.

These are sand dunes of the Sahara Desert in Morocco, Africa. These landforms can shift by wind in desert storms.

Only plants and animals that are suited to a region will grow and live there. For example, only plants and animals that can live in a dry climate with lack of rainfall, will survive. **Deserts** are characterized as getting very little rainfall. There are different kinds of deserts.

There are hot, dry deserts such as the Sahara Desert in North Africa. There are semiarid deserts like in Nevada, USA. There are coastal deserts like the Atacama of Chile. There are also cold deserts which are found in the polar areas of the world, like Antarctica. While some of these desert areas may experience pleasant daily temperatures, and cool nightly temperatures, they all have very little annual rainfall, and therefore are considered desert biomes.

Forests form another major biome of our planet. There are many types of forests. Each has its own climate, plants, and animals. Near the equator, where it is always very hot and rainy, are areas of thick tropical forests, called rainforests.

In the rainforest, the temperature is always above 18ºC and the rainfall is about 250 – 450 mm annually. Rainforests are very humid because of the great amounts of water in the air. In some rainforests, it rains every day. Rainforests have very thick vegetation and most of the trees have broad leaves.

Rainforests are important because they absorb carbon dioxide, a greenhouse gas, and produce oxygen. Rainforests produce nourishing rainfall all around our planet.

OTM2166   ISBN: 9781487710361
© On The Mark Press

Most of the forests that cover North America, Europe, and Asia are **coniferous** and **deciduous forests**. In a coniferous forest, conifer trees such as cedar, fir, and pine trees get their name from the fact that they grow cones which are filled with seeds. Many animals of a coniferous forest feed on these cones. Deciduous trees have trees growing with broad leaves that fall in autumn. Maple, oak, birch, and poplar are some types of deciduous trees.

Grasslands are another of the Earth's biomes. These are large, flat, or gently rolling areas covered with grass. There are many types of grasslands around the world, all with their own kinds of weather.

Prairie grasslands of Alberta, Canada

The grasslands in the cool regions of the world are called temperate grasslands. Temperate grasslands have very few trees because there is not enough rain. In North America these grasslands are called prairies or plains.

Savanna grasslands in Africa

The grasslands in the middle of Eurasia are called steppes. In the middle of the large southern continents, where it is very hot, the grasslands are called savannas.

A savanna is covered in bushes and tough grasses. There are very few trees because most of the year, there is little rain.

The grassland area of South America is called the pampas. It receives very little rainfall and is very dry due to its cold, dry winds.

OTM2166 ISBN: 9781487710361
© On The Mark Press

The top and bottom of the world where there is always snow and ice are the Polar Regions. The climate is always icy cold and there is not vegetation. The animals living there depend mostly on ocean plants and other animals for their food.

Just below the polar areas is the tundra. The **tundra** is characterized as a huge treeless plain. There are three types of tundra worldwide, these being the Arctic tundra, the Antarctic tundra, and the Alpine tundra.

The Arctic tundra, which reaches from the Arctic Ocean to the northern forests, is windy and freezing cold most of the year. The water and land are frozen.

There are only about twenty mammal species living in the Arctic tundra due to the extreme winter temperatures. Most mammals living there must stay active throughout the winter to survive. Hibernation is not an option for them. Only the Arctic ground squirrel is able to hibernate and survive the low winter temperatures in this biome.

Trees will not grow in this area of the world because the earth is permanently frozen beneath the thin active layer (which is only 25 – 100 cm deep). Only when the land thaws a little in the summer can grasses and small shrubs, grow in the active layer of soil. Plants such as mosses, lichens, and grasses grow and provide food for much of the tundra wildlife.

Ecologists also divide the **waters** of the Earth into three biomes. These are the **salt waters**, the **fresh waters**, and the **wetlands**. The salt waters are the oceans and seas. The fresh waters are rivers, lakes, streams, and ponds. The wetlands are bogs, swamps, and marshes.

North Pacific Ocean

Colorado River, Grand Canyon

Wetland area

Researching a Biome

Choose a biome that you would like to learn more about. Access the internet or your local library to find out its location(s) on the Earth, describe the climate, and identify some vegetation and animals that are native to the location you chose to focus on.

Biome: _____

Describe its location(s) on Earth.

_____

_____

_____

Indicate the location(s) of this biome on the world map below, include a legend.

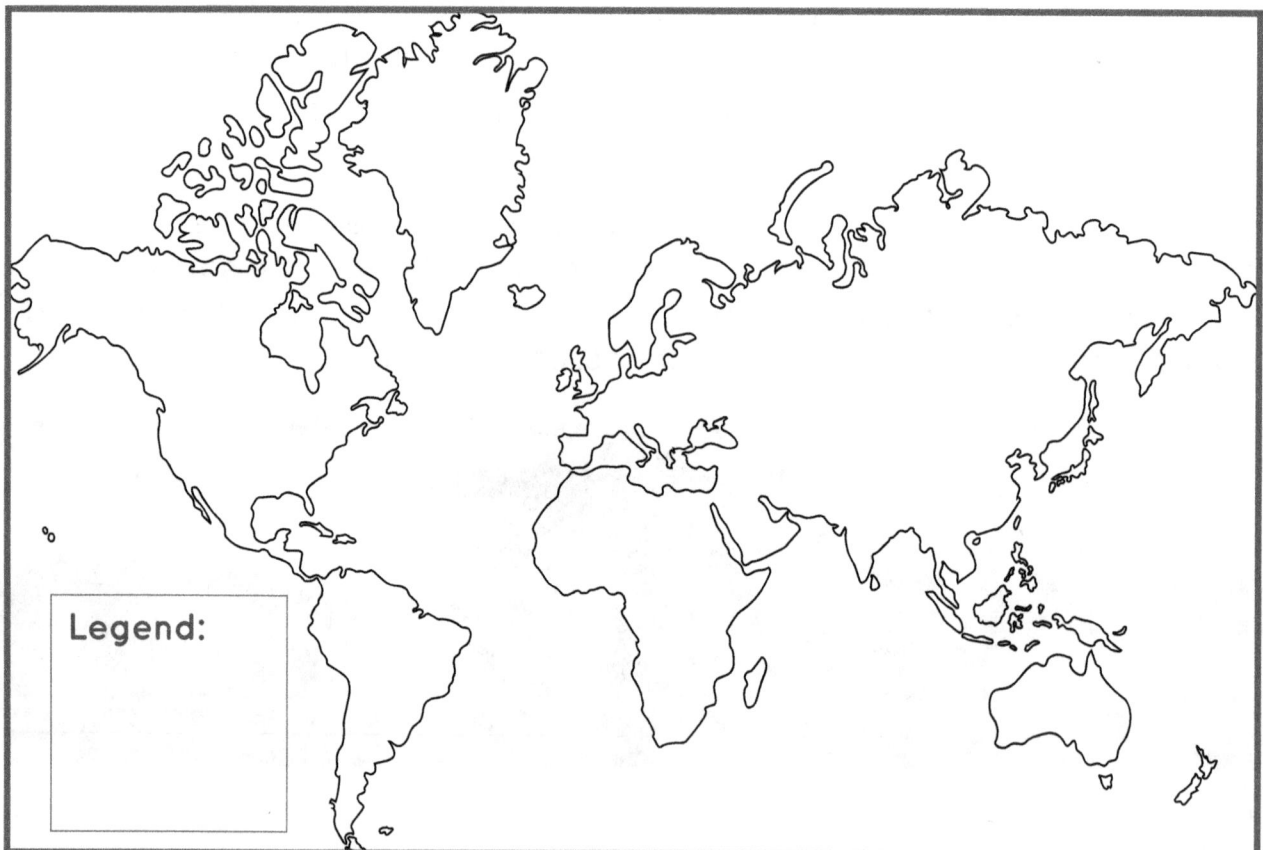

Legend:

OTM2166 ISBN: 9781487710361

Describe the climate in the exact location of the biome that you chose.

_____

_____

_____

_____

_____

Illustrate and label two types of vegetation that can be found in this biome.

[_____]   [_____]

Illustrate and label two kinds of animals that can be found in this biome.

[_____]   [_____]

Name:

FAST FACT!

Research three interesting facts about the biome you chose.

Fact #1

_____

_____

_____

Fact #2

_____

_____

_____

Fact #3

_____

_____

_____

Share your facts with a classmate!

OTM2166   ISBN: 9781487710361

ECOSYSTEMS AT WORK!

LEARNING INTENTION:

Students will learn about biotic and abiotic components and how they interact within ecosystems.

SUCCESS CRITERIA:

- define the meaning of biotic and abiotic components within an ecosystem
- recognize the biotic and abiotic components in an ecosystem
- describe how biotic and abiotic components interact within an ecosystem
- conduct a field study to discover an ecosystem, determine how the components interact
- record observations and conclusions about the components in an ecosystem using diagrams and written descriptions

MATERIALS NEEDED:

- a copy of "Biotic or Abiotic?" worksheet 1 and 2 for each student
- a copy of "Picture This!" worksheet 3, 4, and 5 for each student
- a copy of "Discovering an Ecosystem – A Field Study" worksheet 6, 7, 8, 9, and 10 for each student
- magnifying glasses, small garden shovels, small garden trowels, Popsicle sticks, a few rolls of string, scissors, thermometers, measuring tapes
- a couple of small containers (per group of students)
- access to the internet, clipboards, pencils, pencil crayons, markers, chart paper
- iPods or iPads (*optional*)

PROCEDURE:

*This lesson can be done as one long lesson, or be done in four or five shorter lessons.

1. Give students worksheets 1 and 2 to complete. Read the information to ensure students understand that biotic components are living things, and abiotic components are non-living things within an ecosystem. Both types of components are necessary to make an ecosystem healthy and viable. Students will identify the biotic and abiotic components in the pictures and explain their interaction within an ecosystem.

2. Give students worksheets 3, 4, and 5. They will illustrate the described scenario, identify the biome, describe the climate, identify biotic and abiotic components, explain some interactions, and determine if it is an ecosystem. Students may need to access the internet to research information.

3. Students will be going on a field study to discover an ecosystem. Give them worksheets 6, 7, 8, 9, and 10, a clipboard and pencil, and the materials need to conduct the study. Providing iPods or iPads for photos is an option. Students could conduct the study with a partner or small group. After marking out the area of study, they will draw a detailed map of it, observe and describe objects/animals and the activity, identify biotic and abiotic components, explain some interactions, and determine if there is an ecosystem present.

DIFFERENTIATION:

Slower learners may benefit by working in a small group with teacher support to complete worksheets 3, 4, and 5. Draw the scene on chart paper ahead of time. Proceeding sections could be completed together. This would allow for discussion and review of terminology.

For enrichment, faster learners could choose one biotic component from their field study and research its life cycle and potential threats.

OTM2166   ISBN: 9781487710361
© On The Mark Press

Biotic or Abiotic?

All ecosystems consist of both living and non-living things. Biotic components are living things existing in ecosystems for example, plants and animals. Abiotic components are the non-living things that complete an ecosystem for example, rocks, light, and water.

Identify the biotic and abiotic components of the ecosystems pictured below.

Picture 1

Biotic components:

_____

_____

Abiotic components:

_____

_____

_____

Picture 2

Biotic components:

_____

_____

Abiotic components:

_____

_____

_____

OTM2166 ISBN: 9781487710361
© On The Mark Press

Name:

Biotic and abiotic factors exist in all ecosystems and elements of them are connected in order to make the ecosystem function.

Soil is an abiotic component and a worm is a biotic component in an ecosystem. Can you explain the connection between these two elements?

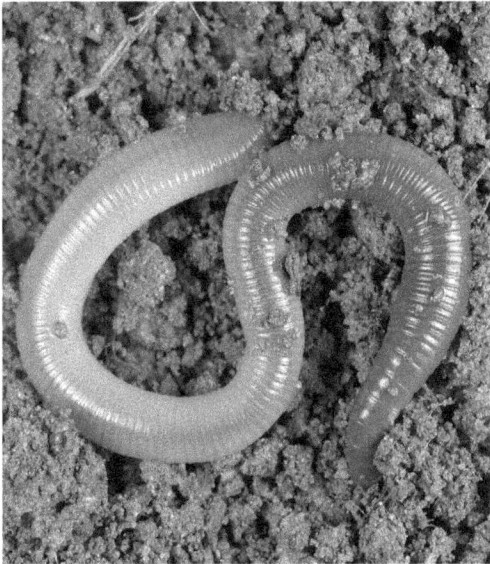

Why does the worm need the soil?

____

____

____

Why does the soil need the worm?

____

____

____

Looking back at the ecosystems pictured on worksheet 1, chose a biotic and an abiotic component. Explain their interaction with each other in their ecosystem.

Picture #1

____

____

____

Picture #2

____

____

____

OTM2166 ISBN: 9781487710361

Picture This!

In the box below, illustrate this scene:

This is the American desert. The sands of the desert are layered in many shades of orange, yellow, and brown. In the foreground, to the left, is a large saguaro cactus. A woodpecker is looking for insects in the cactus. Different kinds of cacti can be seen growing on the desert floor. Several of the cacti plants have brightly colored flowers.

A roadrunner is chasing a rattlesnake. A lizard is hiding in the shade of a large rock in the bottom right hand corner of the picture. A vulture is circling high in the sky, looking for food. Beneath the ground a kangaroo rat is hiding in its burrow. A scorpion is waiting for the rat to come out.

OTM2166   ISBN: 9781487710361
© On The Mark Press

Worksheet 4

Name:

Look back to the scene that you illustrated on worksheet 3. Answer these questions.

What biome of the Earth is it?

Describe the climate of this specific biome location.

List the biotic components from your illustration.

-
-
-
-
-
-
-
-

OTM2166 ISBN: 9781487710361

Look back to the scene that you illustrated on worksheet 3. Complete the following activities.

List the abiotic components from your illustration.

- _____

- _____

- _____

- _____

- _____

- _____

Choose two components and explain their interaction.

_____

_____

_____

_____

_____

Is this an ecosystem? Explain your thinking.

_____

_____

_____

_____

_____

OTM2166 ISBN: 9781487710361
© On The Mark Press

Discovering an Ecosystem – A Field Study

It is time to venture out into your school yard or backyard to explore the systems of living things existing there. Let's discover an ecosystem!

Materials Needed:

- a measuring tape
- a roll of string
- a magnifying glass
- a small garden shovel
- Popsicle sticks

- a small garden trowel
- scissors
- 2 small containers
- thermometer
- an iPod or iPad (*optional*)

What To Do:

1. Chose an area of your yard to field study. Using the string and Popsicle sticks, mark out a square area.

2. On worksheet 7, draw a map of your study location.

3. Investigate your location. Make observations of animals, insects, and vegetation. Make observations of temperature and light. Record your observations in the chart on worksheet 8. *An option is to also take photos of the things you observe.*

4. Take soil samples from two different areas in your study location. Examine the soil content and comment on temperature of each soil sample. Record your observations on worksheet 9.

5. On worksheet 9, make a list of the abiotic and biotic components in your area of study.

6. Make conclusions about the interactions of biotic and abiotic components in your area of study, on worksheet 10.

OTM2166 ISBN: 9781487710361
© On The Mark Press

A map of my chosen area of field study:

OTM2166   ISBN: 9781487710361

Name:

Let's Observe

Record your observations of objects/ factors in the chart below.

Objects/ Factors	Observation/ Activity

Choose an animal or insect that you observed. In your opinion, what is it eating? Justify your answer according to your observations.

_____

_____

_____

_____

Name:

Examine two different soil samples. Comment on each.

Soil sample # 1

_____

_____

_____

Soil sample # 2

_____

_____

_____

List the **abiotic** components in your area of field study.

_____   _____

_____   _____

_____   _____

_____   _____

List the **biotic** components in your area of field study.

_____   _____

_____   _____

_____   _____

_____   _____

OTM2166   ISBN: 9781487710361
© On The Mark Press

Name:

Choose two components that you listed on worksheet 9, explain their interaction.

Let's Conclude

Is there an ecosystem at work in your area of field study? Explain.

Challenge Question:

What could upset the balance in this ecosystem that you identified and examined?

OTM2166 ISBN: 9781487710361

ROLES IN THE NATURAL WORLD

LEARNING INTENTION:

Students will learn about different roles in food chains, and about the balance plants and animals provide in the world's ecosystems.

SUCCESS CRITERIA:

- identify some producers, primary consumers, secondary consumers, and decomposers
- create a composter filled with organic matter
- recognize the role of worms in creating rich soil for plants
- explain the uses and importance of compost to our environment and food chains
- recognize and explain the connection between photosynthesis and the carbon cycle
- conduct an experiment to determine the presence of carbon dioxide
- determine 'true facts' and 'false facts' about living things in the natural world

MATERIALS NEEDED:

- a copy of "It's A Chain Reaction" worksheet 1 and 2 for each student
- a copy of "Roles in the Food Chain" worksheet 3 for each student
- a copy of "Decomposers Hard at Work!" worksheet 4 and 5 for each student
- a copy of "The Carbon Cycle" worksheet 6, 7, 8, and 9 for each student
- a copy of "A Check on Chain Links!" worksheet 10, 11, 12, and 13 for each student
- soil such as sand and loam or topsoil (enough to a classroom composter bin)
- vegetable and fruit scraps, egg shells, leaves, grass clippings, used coffee grinds or tea bags, or other organic material, access to water
- a large plastic bin with a lid, a garden shovel, a long stirring stick, a dozen earthworms
- 2 test tubes with stoppers, 5 boiled lima beans, 5 germinated lima beans, 20 mL of

bromthymol blue, 2 labels (for each pair of students)
- a few test tube racks, a class set of safety glasses
- access to the internet or local library
- chart paper, markers, pencils, pencil crayons

PROCEDURE:

*This lesson can be done as one long lesson, or be divided into six shorter lessons.

1. Using worksheets 1 and 2, do a shared reading activity with the students. This will allow for reading practice and breaking down of the larger words. Along with content, discussion of some vocabulary would be beneficial for comprehension.

   Some interesting vocabulary words to focus on are:

producers	oxygen
molars	secondary consumers
omnivore	photosynthesis
carbon dioxide	chloroplasts
primary consumers	cellular respiration
herbivores	minerals
predator	carnivores
molecules	

2. Divide students into pairs, give them worksheet 3. They will engage in a 'think-pair-share' activity to discuss and record some examples of producers, primary consumers, secondary consumers, and decomposers. Some students may require access to the internet. Follow up by coming back together as a large group to record students' responses for each category on large chart paper, this can be posted in the classroom. Discuss the meaning of tertiary consumers (a carnivore at the **top** of a food chain that feeds on primary and secondary consumers). Ask students to review their list of secondary consumers and circle any tertiary consumers.

OTM2166   ISBN: 9781487710361
© On The Mark Press

3. Working as a large group, students will create and contribute to a classroom composter. Give students worksheet 4. Read through the materials needed and what to do sections to ensure students' understanding. Gather your materials and start building! Give students worksheet 5. They will illustrate the composting layers, explain how they will use the compost, and give reasons for composting. Emphasis should be made on the positive effects composting has on our soils and for our environment. Also, composting enriches the soils for our plants which are part of our food chain.

4. Give students worksheet 6. Read through the information with students to ensure their understanding.

5. Explain to students that they will experiment with cellular respiration. Divide students into pairs or small groups and give them worksheets 7, 8, and 9, and the materials to conduct the experiment. (Students should conclude that cellular respiration involves taking sugar and breaking it down. Seeds of plants are full of sugar. Breaking down sugar releases carbon dioxide. This allows the plant to harvest energy to produce roots, stems, and leaves. This is the same concept for humans as we break down the sugar molecules we get from eating food, energy is released for us to use to move and grow. When we break down sugar molecules, carbon dioxide is released, just as it was in the plant experiment.)

6. Give students worksheets 10, 11, 12, and 13 to complete.

DIFFERENTIATION:

Slower learners may benefit by:

- completing worksheets 10 and 11 with a strong peer
- working in small group with teacher direction to complete worksheets 12 and 13, answers could be recorded on chart paper, then later shared with the large group as a check- up/ follow up

For enrichment, faster learners could design a garden space (wherein the compost will be used).

It's a Chain Reaction

All living things need energy to grow. Without energy, you and all other living things would not survive. What is energy? Where do all living things get the energy they need?

Energy comes from the sun in the form of heat and light. The only living things that can trap the sun's energy are plants. Plants store energy for all living things to use.

Everything that we eat started with energy from the sun which was trapped by plants.

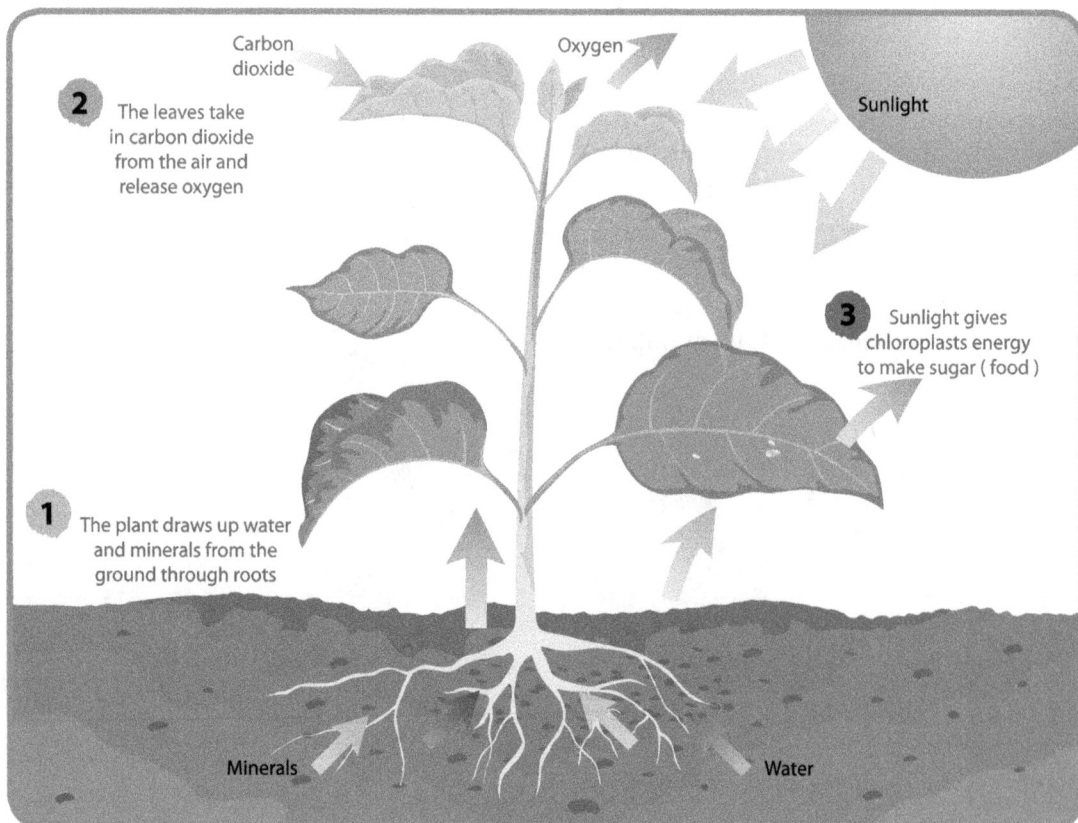

Without plants, there wouldn't be any life on Earth. We get the energy we need to survive from the food that we eat. The energy from plants is passed to all other living things through food chains.

Everything we eat is part of a food chain. Plants are **producers** because they trap the sun's energy and use it to make their own food, in a process called photosynthesis.

Carbon dioxide

Oxygen

Sunlight

2 The leaves take in carbon dioxide from the air and release oxygen

3 Sunlight gives chloroplasts energy to make sugar (food)

1 The plant draws up water and minerals from the ground through roots

Minerals

Water

OTM2166   ISBN: 9781487710361
© On The Mark Press

Plants then provide food for other living things. Some animals are **herbivores** because they eat only plants. The energy from plants is passed on to herbivores that eat them.

Herbivores are called primary consumers because they get energy directly from the plants. A deer, a rhino, and a turtle are examples of herbivores. Did you know that many herbivores have wide molars that help them grind up leaves and grasses?

Rabbits are also primary consumers because they only eat plants. When a rabbit eats a plant, the energy in the plant is passed on to the rabbit. This energy helps the rabbit grow and reproduce. When the rabbit is caught and eaten by a **predator** such as an owl, the energy that is contained in the rabbit is then passed on to the predator. Animals like the owl, which eat herbivores, are called **secondary consumers**. They are also called **carnivores**. Carnivores are animals that eat only meat.

An **omnivore** is a living creature that eats both plants and animals. Most humans eat both plants and animals, so we are considered omnivores.

Many carnivores eat herbivores, but some eat omnivores and other carnivores. A crocodile, an eagle, and a lion are examples of carnivores. Did you know that many carnivores have long sharp teeth and claws that help them grab prey and eat its' meat?

OTM2166   ISBN: 9781487710361

Roles in the Food Chain

With a partner, discuss your ideas about producers, primary consumers, secondary consumers, and decomposers. Record some examples of each type.

Producers:

Primary Consumers:

Secondary Consumers:

Decomposers:

OTM2166   ISBN: 9781487710361
© On The Mark Press

Decomposers Hard at Work!

Worms are decomposers. They eat vegetation then deposit castings in soil to make it rich in nutrients. This soil can be used in gardens to help grow healthy plants.

Some gardeners work to get this rich soil by composting. Composting is a process that breaks down food waste, leaves, grass clippings, and wood bits into humus. Adding in some decomposers like earthworms is the secret ingredient! Let's try composting!

Materials Needed:

- a large plastic bin with a lid
- a drill
- garden soil
- vegetable and fruit scraps
- egg shells
- about 10 earthworms
- leaves, grass clippings
- used coffee grinds or tea bags
- a garden shovel
- access to water
- a long stick for stirring

What To Do:

1. Put some soil in the bottom of the plastic bin.
2. Add some vegetable or fruit scraps, egg shells, leaves, grass clippings, and used coffee grinds or tea bags on top of the soil.
3. Moisten the soil with a little bit of water.
4. Repeat steps 1, 2, and 3 to make layers. Add in the earthworms.
5. Complete worksheet 8.
6. Using the drill, *your teacher* will make some holes in the lid of the bin. Put the lid on the bin to close it, and place it outside.
7. Use the long stick to stir your compost pile each week. Add organic materials when available. After several weeks, you will have created some humus that you can add to a garden to grow plants.

Draw a diagram of the composter. Show the layers of ingredients you put in it.

Food for Thought!

How will you use your compost?

_____

_____

_____

List some reasons why people should compost.

• _____

• _____

• _____

• _____

• _____

• _____

OTM2166   ISBN: 9781487710361
© On The Mark Press

The Carbon Cycle

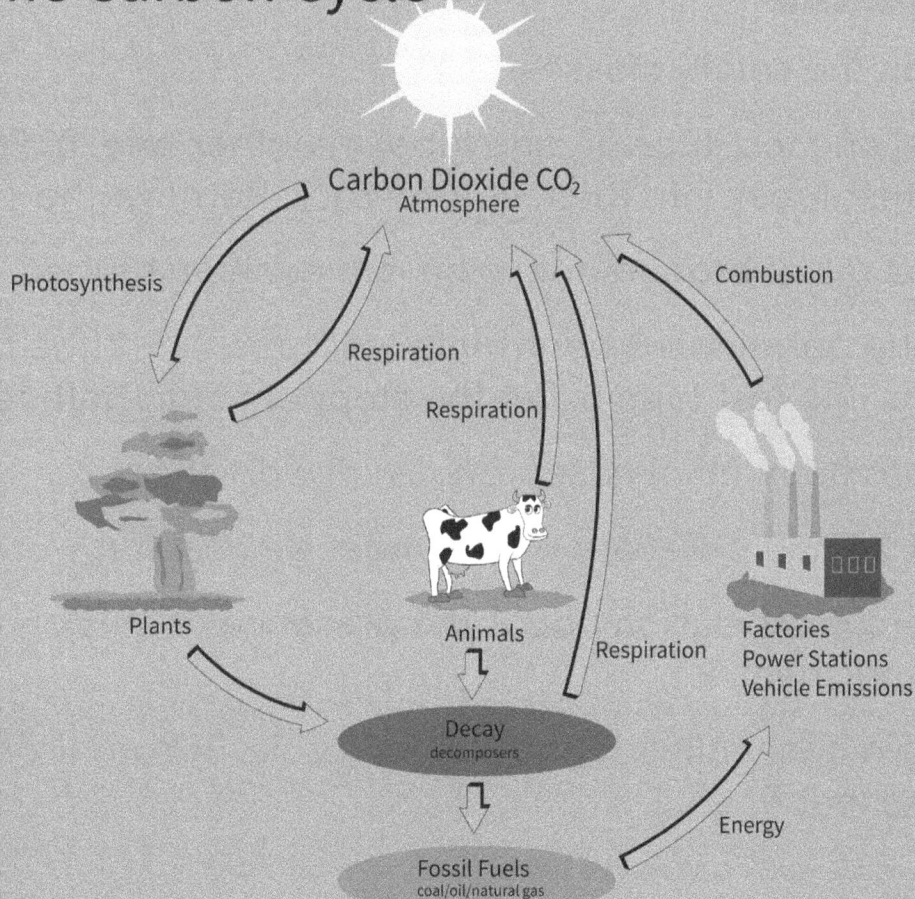

The energy that living things get from eating food is achieved through a process called cellular respiration. During cellular respiration, the oxygen in the cells of living things breaks down the sugar molecules obtained from eating food, and releases energy for living things to use to move and grow. As the sugar molecules are broken down, carbon dioxide is released.

Plants use this carbon dioxide, along with water and light energy from the sun to create chlorophyll. Chlorophyll is what is needed to make sugar food for plants. Plants release oxygen into the air, which animals and humans breathe in.

Photosynthesis and cellular respiration are part of the cycling of matter and the transfer of energy in the world's ecosystems.

The Carbon Cycle

Carbon Dioxide CO_2
Atmosphere

Photosynthesis

Combustion

Respiration

Respiration

Plants

Animals

Respiration

Factories
Power Stations
Vehicle Emissions

Decay
decomposers

Energy

Fossil Fuels
coal/oil/natural gas

OTM2166 ISBN: 9781487710361
© On The Mark Press

Try this!

Let's experiment with bean seeds to test for carbon dioxide. Is this by-product present as the living thing, a germinated bean seed, uses its sugar molecules to break down the food that it makes for itself?

Materials Needed:

- 2 test tubes with stoppers
- 5 boiled lima beans
- 5 germinated lima beans
- 2 labels

- 20 mL of bromthymol blue
- a test tube rack
- a pair of safety glasses
- a marker

What To Do:

1. Put on the safety glasses.

2. Label one test tube 'A', and label the other one 'B'. Place the two test tubes into the test tube rack. Remove the stoppers.

3. Put 10 mL of bromthymol blue into each test tube.

4. Put the germinated beans into test tube A. Put the boiled beans into test tube B. Put the stoppers onto both test tubes.

5. Record your initial observations on worksheet 8.

6. Wait 24 hours. Record the changes you see on worksheet 8.

7. Record a conclusion about the presence of carbon dioxide on worksheet 9.

8. Make a connection to the carbon cycle. Record it on worksheet 9.

OTM2166   ISBN: 9781487710361
© On The Mark Press

Name:

Let's Observe

This is a diagram of the beans in the test tubes at the start of the experiment:

The color of the liquid in test tube A is _____.

The color of the liquid in test tube B is _____.

This is a diagram of the beans in the test tubes after 24 hours:

The color of the liquid in test tube A is _____.

The color of the liquid in test tube B is _____.

Observation tip!

Carbon dioxide is colorless. The bromthymol blue acts to indicate its presence.

Let's Conclude

In which test tube was the presence of carbon dioxide observed? Explain.

_____

_____

_____

_____

Explain why there was a presence of carbon dioxide.

_____

_____

_____

_____

Support your thinking with a diagram or formula:

OTM2166   ISBN: 9781487710361
© On The Mark Press

A Check on Chain Links!

Write T for 'true' and F for 'false' for each of the following sentences.

_____ a)　Only humans need energy to live.

_____ b)　There wouldn't be any life on Earth without plants.

_____ c)　Humans get energy directly from the sun.

_____ d)　All living things can trap the sun's energy.

_____ e)　Plants are called primary consumers.

_____ f)　Plants pass on energy through food chains.

_____ g)　An owl is a secondary producer.

_____ h)　Herbivores are primary consumers.

_____ i)　Animals that eat herbivores are secondary consumers.

_____ j)　Carnivores eat plants and animals.

_____ k)　Predators eat plants.

_____ l)　Insectivores eat insects.

_____ m)　Humans that eat both plants and animals are omnivores.

_____ n)　Bees are primary producers.

_____ o)　Fungi are decomposers.

_____ p)　Worms are omnivores.

_____ q)　Through cellular respiration, carbon dioxide is released.

Name:

Fill in the blanks with the words in the Word Box.

energy	producers	heat	decomposers	food chains
sun	leaves	roots	omnivores	secondary consumers
stems	seeds	light	fruit	primary consumers

At the beginning of every food chain is the _____.
All living things need energy to survive. They obtain
energy directly or indirectly from the sun. The sun
provides energy in the form of _____ and
_____. Without the sun, nothing could survive.

Plants need _____ from the sun to grow.
Plants are called _____ because they
produce materials that can be eaten by other living
things. Some parts of plants that are consumed by other
living things are _____, _____
, _____, _____, and
_____.

Living things that rely on plants only for their food
source are called _____. Living things that
hunt prey for their food source are considered to be
_____. _____ are living things that
have a meat and plant based diet.

_____ eat dead plants and animals and
release chemical nutrients back into soil, water, or air.

Plants, animals, and humans are all part of
_____.

OTM2166   ISBN: 9781487710361
© On The Mark Press

Name:

Photosynthesis is a system. There is an input, a process, and an output. Together, they perform a function, working for a common goal.

Process of Photosynthesis

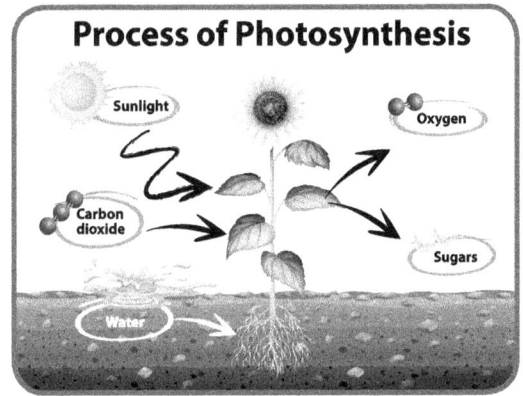

Use this word equation to answer the following questions about photosynthesis:

Water + carbon dioxide + light energy → chlorophyll → sugar + oxygen

1. What are the inputs?

   _____

   _____

2. Describe the process that takes place.

   _____

   _____

3. What are the outputs?

   _____

   _____

   _____

4. How does the output of this system impact the environment?

   _____

   _____

   _____

Cellular respiration is a system. There is an input, a process, and an output. Together, they perform a function, working for a common goal.

Use this word equation to answer the following questions about cellular respiration:

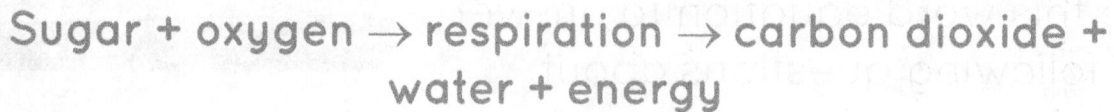

Sugar + oxygen → respiration → carbon dioxide + water + energy

1. What are the inputs?

_____

_____

_____

2. Describe the process that takes place.

_____

_____

_____

3. What are the outputs?

_____

_____

_____

4. How does the output of this system impact the environment?

_____

_____

_____

OTM2166   ISBN: 9781487710361
© On The Mark Press

FOOD CHAINS AND WEBS

LEARNING INTENTION:

Students will learn about and build food chains and food webs consisting of different living things.

SUCCESS CRITERIA:

- describe the connection of living things, beginning with the sun
- create food chains using pictures and written descriptions
- create a food web using pictures and written descriptions
- identify biotic and abiotic components within an ecosystem, explain their interdependence
- determine the producers, primary consumers, secondary consumers, and decomposers existing within a particular ecosystem
- create a food web to illustrate how animals live in balance with one another in an ecosystem
- record factors that could affect the balance of nature in an ecosystem

MATERIALS NEEDED:

- a copy of "Linking It Together" worksheet 1 and 2 for each student
- a copy of "Make a Food Web" worksheet 3 and 4 for each student
- a copy of "A Food Web Challenge!" worksheet 5 for each student
- a copy of "Wolf Island" worksheet 6, 7, and 8 for each student
- access to the internet or local library
- a read aloud (see #5 in procedure section)
- different colored tokens or cubes, pinnies
- glue, scissors, clipboards, chart paper, markers, pencils, pencil crayons

PROCEDURE:

*This lesson can be done as one long lesson, or be done in five or six shorter lessons.

1. Give students worksheet 1. Read the information and discuss to ensure students' understanding of the material. Give students worksheet 2 to complete. Students may need access to the internet or a local library to get information to complete their food chain. *A follow-up option is to have students present their food chain within a small group. This will enhance their knowledge of different food chains that exist on our Earth, and reinforce the understanding that every living thing has a purpose.*

2. Broaden students' knowledge by discussing food webs. Explain that all plants and animals are interconnected. Many connections exist within a small group of plants and animals. Using worksheets 3 and 4, students will create a food web.

3. Give students worksheet 5. They may need access to the internet or a local library to get information to complete this food web challenge. A follow up option is to have students explain their food web in words, and then attach it to their food web display. The food webs could be displayed on a bulletin board.

*As a further activity to enhance the learning about food chains, show students Bill Nye the Science Guy episode called "Food Web". Accessing the internet, go to www.youtube.com for the full episode.

4. Play the 'Lap Sit Game'. Have students form a **tight** circle, with each student facing forward, looking at a peer in front of them. Once students are still, instruct them to slowly sit down, resting upon the peer behind them. If done correctly, they will support each other in a sitting position. Upon success of this activity, ask students to stand. Pose these questions:

- Why were you all able to sit without falling?
- How can this activity be related to a food chain?
- What would happen if one person in the circle fell?
- What would happen if one species in a food chain was not available?
- Would or could this food chain sustain?

5. As a further reinforcement of the concepts discussed, a suggested read aloud is <u>Wolf Island</u> (Author: Celia Godkin). After reading, engage students in a discussion of how the natural balance in the island's ecosystem was affected by the absence, and the return, of the wolf species. Give students worksheets 6, 7, and 8 to complete.

6. Play an "Ecosystem Tag" game, where students are assigned to be different animals in your local area. Students can also hunt for different colored tokens that represent components of survival in that area (e.g., food, water, shelter, reproduction). Introduction of a negative impact to the living things in the area (e.g., weather systems, drought, locusts, hunters, etc.), can also be done to make the game more interesting. The impact may only affect a certain species, which could upset the natural balance in the area. Pause the game every so often to tally the number of each species still alive. Engage students in a discussion about the reasons for the disturbance in the natural balance, and what else/ who else this could impact. Pose these questions:

- What things could ultimately affect nature's balance in an ecosystem?
- What effect could the introduction of a new plant or animal species have on an ecosystem?
- Do changes to the environment have more impact on a specialized species rather than generalized species?

DIFFERENTIATION:

Slower learners may benefit by:

- working with a partner to make a food web on worksheet 5
- working together as a small group with teacher direction to complete worksheets 6, 7, and 8; alternatively, these learners could be given a reduction in expectations by only completing portions of worksheets 6, 7, and 8

For enrichment, faster learners could:

- work as a small group, choosing one of their food web challenge creations, to create a skit/ short play to demonstrate how the living things in this food web are connected. They could perform this skit for the rest of their classmates and/or for students in other classrooms.
- investigate any special adaptations that some living things have that help them to exist in the ecosystem on Wolf Island. These could be presented to the large group to engage students in a discussion about adaptations.

OTM2166   ISBN: 9781487710361
© On The Mark Press

Linking It Together

The energy that starts with the sun is passed from one living thing to another through food chains. Animals get energy from plants and other animals because they provide food for each other. They are all part of a food chain.

So, it all begins with the sun. The sun's heat energy makes plants grow. Plants are food for some animals. Some animals are food for each other.

This is a food chain. Grass is food for a grasshopper. A grasshopper is food for a mouse. A mouse is food for a snake. A snake is food for a hawk.

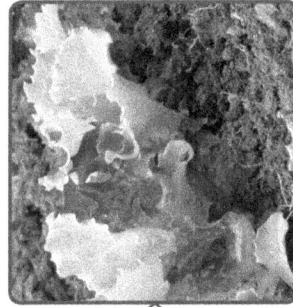

Name:

Create two food chains. Start with the sun!

Explain this food chain:

Explain this food chain:

OTM2166   ISBN: 9781487710361
© On The Mark Press

Make a Food Web

Food chains are all connected. A food web is a bunch of food chains. Let's create a food web!

Materials Needed:

- scissors

- glue

What To Do:

1. Cut out the pictures on worksheet 3.

2. On worksheet 4, glue the pictures to make a food web.

3. Draw lines between plants or animals that eat each other.

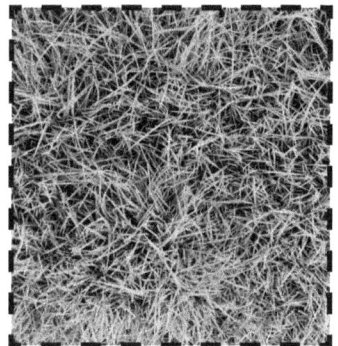

Name:

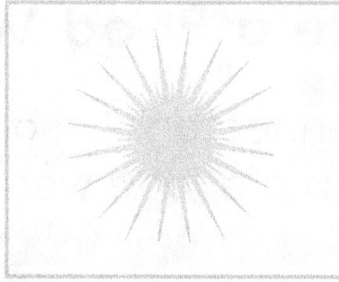

Why is it called a food web and not a food chain?

_____

_____

_____

_____

OTM2166   ISBN: 9781487710361
© On The Mark Press

Name:

A Food Web Challenge!

Create your own food web. In the boxes, draw pictures of biotic things that are connected in a food web. To complete it, draw lines between plants or animals that eat each other. Describe your food web to a classmate.

Wolf Island

List some of the biotic components on Wolf Island.

- _____
- _____
- _____
- _____

- _____
- _____
- _____
- _____

List some of the abiotic components on Wolf Island.

- _____
- _____
- _____
- _____

- _____
- _____
- _____
- _____

Choose one biotic and one abiotic component that you listed and explain their interdependence within the ecosystem on Wolf Island.

_____

_____

_____

Provide another example of biotic and abiotic interdependence within the ecosystem on Wolf Island.

_____

_____

_____

OTM2166 ISBN: 9781487710361
© On The Mark Press

Worksheet 7 Name:

Sort the biotic components that you listed on Worksheet 6 into the following categories:

Producers	Primary Consumers	Secondary Consumers	Decomposers

What happened to the island ecosystem when the wolves left?

- ____
  ____
- ____
  ____
- ____
  ____
- ____
  ____
- ____
  ____
- ____
  ____

Name:

Draw a food web to display how the animals on Wolf Island lived interdependently.

What could happen if there is too much or too little of a particular life species in an ecosystem?

_____

_____

_____

Does a habitat have a limit to the amount of living things that can live in it? Explain.

_____

_____

_____

_____

OTM2166   ISBN: 9781487710361
© On The Mark Press

ECOLOGICAL SUCCESSION

LEARNING INTENTION:

Students will learn about primary and secondary succession; how to create an ecosystem that is able to sustain and grow.

SUCCESS CRITERIA:

- create a terrarium for an ever changing ecosystem to exist in
- document changes in the terrarium over time, using diagrams and written descriptions
- explain how the water cycle works to provide moisture for soils and plants
- describe primary and secondary succession, giving examples of its occurrence
- research and record examples of how animals/insects impact succession

MATERIALS NEEDED:

- a copy of "An Ever Changing Ecosystem" worksheet 1, 2, and 3 for each student
- a copy of "Succession" worksheet 4, 5, and 6 for each student
- a 2L clear plastic pop bottle, a small gardening shovel, a few decorative rocks, a few small plants (e.g., ferns, moss, ivy, begonias) for each group of students
- terrarium gravel, activated garden charcoal, potting soil, newspaper, a few spray bottles
- access to water
- access to the internet or library, clipboards, pencils, pencil crayons, markers, chart paper

PROCEDURE:

*This lesson can be done as one long lesson, or be done in three or four shorter lessons.

1. Give each student worksheets 1, 2, and 3. Divide students into small groups. Give each group the materials to create a terrarium. After creating a terrarium, students will draw a diagram to depict how it is layered and what items are in it.

*Alternatively, this activity could be done with one large group to create only one classroom terrarium.

2. Using worksheets 4 and 5, do a shared reading activity with the students. Along with content, discussion of some vocabulary would be beneficial for comprehension.

   Some interesting vocabulary words to focus on are:

primary succession	algae
micro-organisms	germination
secondary succession	chlorophyll
sustained	vegetation
climax community	fungi
rejuvenation	habitat

3. Give students worksheet 6, and a clipboard and pencil. With access to the internet or local library, they will research the roles animals and insects have in succession. A follow-up option, is to have students share one of their examples with a partner or within a small group.

DIFFERENTIATION:

Slower learners may benefit by only researching and recording one example of how animals/ insects participate in ecological succession on worksheet 6.

For enrichment, faster learners could:

- research how forest fires can become so intense that it creates its own weather patterns
- research to learn more about 'controlled burns' and how they are used in forestry/ land development
- do some research or interview an Aboriginal elder to learn about the Aboriginal perspective on sustainability, and the ways Aboriginal people manage habitat and wildlife

An Ever Changing Ecosystem

A terrarium is like a mini greenhouse that houses and grows an ecosystem inside it. It is created in a clear container so that sunlight can enter and warm the air, soil, and plants that are growing inside it. Let's get started on creating a terrarium!

Materials Needed:

- 2L plastic pop bottle with cap
- terrarium gravel
- potting soil
- activated charcoal
- a spray bottle
- scissors

- a small gardening shovel
- water
- small plants (e.g., ferns, moss)
- a marker
- newspaper
- a few small decorative rocks (*optional*)

What To Do:

1. Spread some newspaper onto a table top. Use this as your work space.
2. Mark a small 'X' on the side of the plastic pop bottle, about half way down. Draw a line from the 'X' around the circumference of the bottle. Using the scissors, carefully make a hole on the 'X' and cut along the line you have drawn, to cut the bottle into two parts.
3. Place about a 4 cm layer of terrarium gravel into the bottom portion of the pop bottle.
4. Add about a 1 cm layer of activated charcoal on top of the gravel layer.
5. Fill the rest of the bottom portion of the pop bottle with potting soil. Mist with water from the spray bottle if it is dry (just to moisten).
6. Plant small plants into the terrarium and a few decorative rocks. Be sure there is enough space between them so that the plants have room to grow.
7. Place your terrarium in indirect light. If necessary, use the spray bottle to mist your plants with water, only as needed to keep the soil moist.
8. Watch your plants grow! Complete worksheets 2 and 3 as you observe over time.

OTM2166   ISBN: 9781487710361
© On The Mark Press

Name:

This is a diagram of my terrarium at the **start** of the project:

This is a diagram of my terrarium after ______________ weeks/months.

Describe the changes that occurred over time in the ecosystem that you created.

____

____

____

____

OTM2166 ISBN: 9781487710361

Name:

Let's Connect It!

Did you know that water moves through ecosystems in a cycle? This is called the water cycle. Refer to the water cycle diagram below to learn more.

As the sun heats the ground, oceans, rivers, lakes, and streams, the water evaporates. Evaporation is water in its gas form rising again to condense as a cloud.

Condensation happens when water in a gas form, meets cooler air. It will form a cloud in the sky where this gas changes into droplets of water.

Water cycle

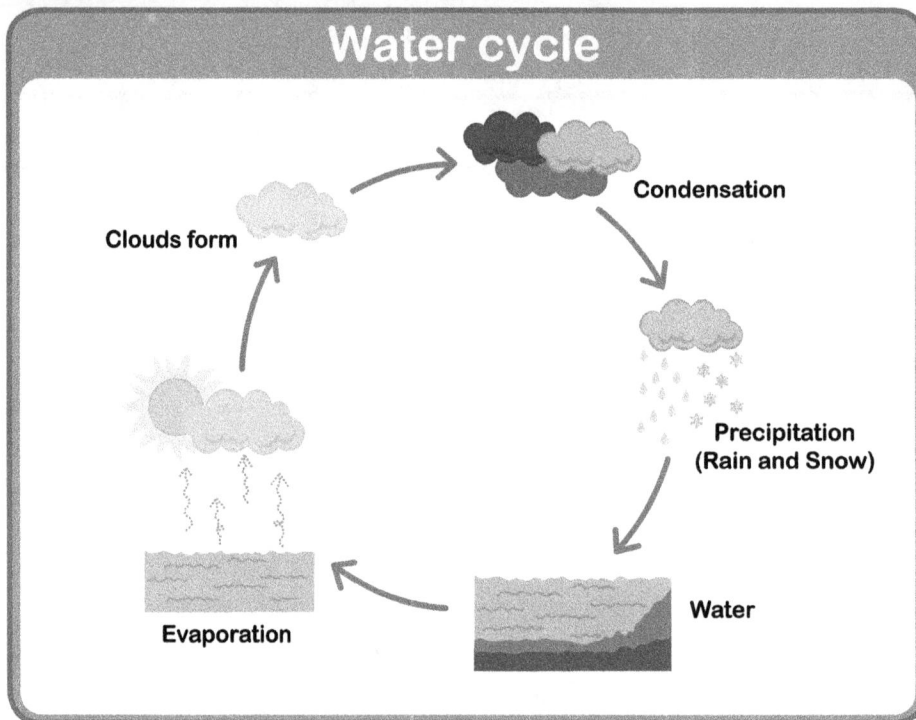

Clouds form

Condensation

Precipitation
(Rain and Snow)

Water

Evaporation

These droplets of water, called precipitation, fall back down to Earth. Precipitation can be in the form of rain, snow, sleet, or hail. It wets the ground, and fills up rivers, lakes, oceans, and streams.

Explain how the atmosphere and moisture content in your terrarium is similar to the conditions of the atmosphere and moisture content on our planet Earth.

_____

_____

_____

OTM2166   ISBN: 9781487710361
© On The Mark Press

Name:

Succession

You have learned that ecosystems change over time. This change is known as succession. Succession can mean new growth, and replacement of old. There are two types of succession, primary succession and secondary succession.

Primary succession is when a **new** community is created. This may happen in different ways and in different places.

An example of a new community may begin with the eruption of a volcano, where the lava spewed, cools and hardens to form a new landscape.

Some organisms that would be first to appear on this new landscape are ones that can secure themselves to rock, like lichens. Lichens are algae and fungi. The algae contains chlorophyll and can make food, the fungi absorbs water easily. They work together to survive and grow.

The fungi and algae in lichen have an important interaction. Fungus in lichen protects it from the environment and from loss of moisture. The algae in lichen provide the food source to grow. When the fungus is fed, it has the strength to break down minerals in the rock around it, to create soil for new plants to grow in.

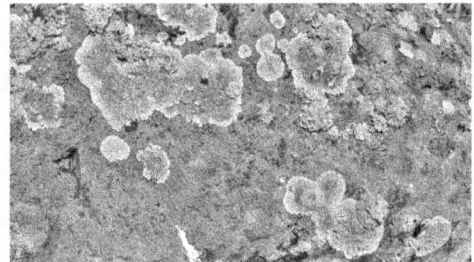

The fungi in lichen break down the minerals in rock, and soil is created. In a minimum amount of soil, mosses can grow. As more lichen and mosses are produced and die off, more soil is made. When there is enough soil, grasses and weeds will begin to grow.

Over time, shrubs may appear and even trees. At this stage, it is considered a **climax community**. Throughout the growth of vegetation, different animals and micro-organisms appear. New ecosystems are created and there exists a lot of diversity. The community is sustained.

OTM2166 ISBN: 9781487710361
© On The Mark Press

Secondary succession happens after an existing community is partially or completely destroyed. This destruction or damage may occur due to forest fires, extreme weather incidences such as tornados, floods, or just simply because the land was unattended for extended amount of time. In secondary succession, the damaged community is eventually replaced by other communities.

Forest fires can be devastating. It can completely destroy a habitat to many animals, vegetation, and even the tiniest of organisms existing within it.

But, with destruction, there becomes room for new growth. There is rejuvenation. Slowly the blackened area turns green again as seedlings germinate and new, young vegetation appears. This is secondary succession.

As the young vegetation matures and spreads in the new forest area, it attracts more and more animals, birds, and micro-organisms.

Often, these areas will grow to be even fuller and healthier than the pre-existing forest. This provides an even richer habitat for ecosystems in the forest to survive in.

Let's Connect It!

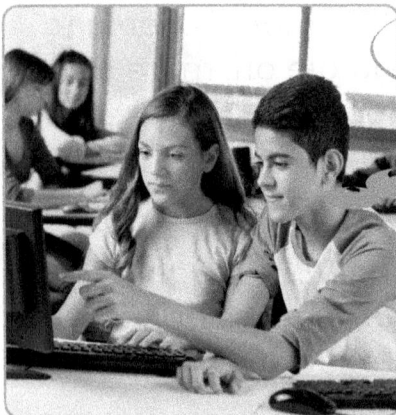

Discuss with a partner some examples of secondary succession. In what instances/places could secondary succession happen?

OTM2166   ISBN: 9781487710361
© On The Mark Press

Name:

Research It!

Did you know that insects and animals have an important role in natural succession? Do some research to find out how they help to build up a primary ecosystem, or to re-build a secondary ecosystem. Provide three examples.

Example #1

_____

_____

_____

_____

Example #2

_____

_____

_____

_____

Example #3

_____

_____

_____

_____

OTM2166   ISBN: 9781487710361
© On The Mark Press

HUMAN ACTIONS AND TECHNOLOGY

LEARNING INTENTION:

Students will learn about the impact human actions and technologies have on ecosystems in the environment, and the organizations that are working to make a difference.

SUCCESS CRITERIA:

- identify human actions that alter balances and interactions in the environment
- identify the technologies that alter balances and interactions in the environment
- determine if human activities or technology is having a positive or negative impact
- create two dioramas that respectively depict the 'before' and 'after' of human and/ or technological interference
- recognize the costs and benefits of human and technological involvement
- research, record, and share information about an ecosystem that is being affected by human activity or technology
- research, record and share information about an organization that works toward ecological sustainability
- create a pamphlet to inform the public about the efforts of the organization

MATERIALS NEEDED:

*ask students to bring in two shoe boxes/ cardboard boxes each for diorama creations

- a copy of "Altering the Balance" worksheet 1, 2, and 3 for each student
- a copy of "Before and After" worksheet 4 and 5 for each student
- a copy of "Making an Impact" worksheet 6 for each student
- a copy of "Taking Action!" worksheet 7, 8, and 9 for each student
- access to the internet or local library
- Plasticene or modeling clay, construction

paper, glue, scissors, masking tape, packing or duct tape, assorted paints and paint brushes, pipe cleaners, Styrofoam trays, paper plates and cups, small pieces of fabric, or any materials suited for creating dioramas

- a selection of magazine or newspaper articles that feature stories on human or technological impacts on an ecosystem
- clipboards, pencils, pencil crayons, scissors, markers, chart paper
- iPads or Macbooks (optional materials)

PROCEDURE:

*This lesson can be done as one long lesson, or be divided into seven shorter lessons.

1. Divide students into pairs and give them worksheet 1. They will engage in a 'think-pair-share' activity to discuss and answer the questions. Encourage students to think of not only negative ways, but positive ways also. A follow up option is to come back together as a large group to hear some responses. Record them on chart paper for future reference.

2. Give students worksheets 2 and 3. Students will comment on different environmental situations that have perhaps experienced human or technological impact. Read through and discuss the example to ensure their understanding of the task. Some students may need to access the internet to gather and consider information.

3. Give students worksheet 4. Read through the assigned task instructions with students to ensure their understanding. After choosing an environmental situation they would like to feature, they will plan what it should look like before human activity or technology impact, and plan what will look like after impact. (This activity and diorama creation could be done individually or in pairs.) Once students have completed their plan, give them the materials to begin creating their two dioramas.

OTM2166 ISBN: 9781487710361
© On The Mark Press

4. Upon completion of their two dioramas, give students worksheet 5 to complete and attach to their displays. Dioramas could be displayed in the classroom or in a special location around the school for others to see and learn more about.

5. Have students chose a newspaper or magazine article that features/ informs about human actions or technology on an ecosystem. After reading it, give them worksheet 6 to complete. They will share what they have learned with a classmate.

6. Give students worksheets 7, 8, and 9. With access to the internet, students will research an organization that works on the principles of sustainable development. They will comment what the organization does and why, using information from their website or articles to support their answer. They will also explain how they can contribute to the cause. Students will share what they have learned with a classmate.

7. At the bottom of worksheet 8, students will brainstorm ideas to be included in an informational pamphlet about the organization's focus, their efforts to help the cause, and how others can get involved. The pamphlet will be created on worksheet 9. A follow up option is to have students cut out their pamphlets and fold them along the dotted lines. A bulletin board display could be created using these pamphlets.

DIFFERENTIATION:

Slower learners may benefit by:

- working as a small group with teacher direction to complete worksheets 2 and 3, this would allow for rich discussion to occur and an opportunity to use related scientific vocabulary

- working as a small group with teacher support to read through an article that informs about examples of human actions or technology on an ecosystem, after some discussion about the article, worksheet 6 could be completed together on chart paper

- working with a strong peer to research about the efforts of an ecological organization, completing worksheet 7 together

For enrichment, faster learners could:

- create a multimedia presentation to outline the level of damage and stage of recovery of the environmental situation that they portrayed in their dioramas (worksheets 4 and 5)

- create a multimedia presentation to outline the level of damage and stage of recovery in accordance to the information in the article that they read about (worksheet 6)

- create an informational video to promote the cause and efforts of the ecological organization that they researched about

- write a letter to government officials explaining the concerns and possible ways to alleviate the problems that were outlined by the ecological organization that they researched

Altering the Balance

Think **Pair** **Share**

With your partner, do some thinking and sharing of ideas about the questions below.

"What human actions or activities alter balances and interactions in the environment?"

Record your thinking.

"What technologies alter or impact balances and interactions in the environment?"

Record your thinking.

OTM2166   ISBN: 9781487710361

For the following human and technological impacts, comment on whether it is a positive impact, a negative impact, or both on the environment. Be sure to explain your thinking.

Example:

Deforestation

This has a negative impact. When trees are cut down, they can no longer create oxygen for humans to breathe in. Some animals lose their homes. This can impact the balance in forest food chains.

Urbanization

_____

_____

_____

Water Treatment

_____

_____

_____

Fishing

_____

_____

_____

Continue to comment on human and technological impacts. Are they a positive impact, a negative impact, or both on the environment? Be sure to explain your thinking.

Recycling

_____

_____

_____

Mining

_____

_____

_____

Tree Planting/Green Belt Development

_____

_____

_____

Hunting

_____

_____

_____

_____

OTM2166   ISBN: 9781487710361
© On The Mark Press

Before and After

Choose an environmental situation that human activity or technology has impacted, either from the ones listed on Worksheets 2 and 3, or by using an idea of your own.

Your task will be to create *two* dioramas to display in detail the 'before' and 'after' the impact of human activities or technologies have had on this environmental situation.

Plan your display:

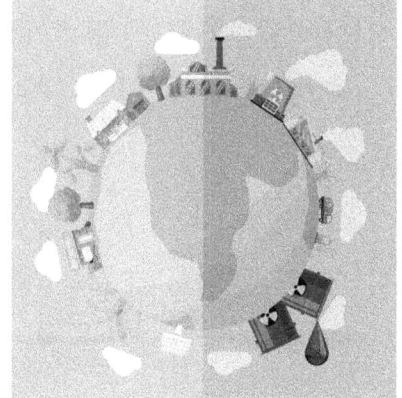

What will you include in diorama #1? (before the impact)

What will you include in diorama #2? (after the impact)

Now gather your materials and begin creating your diorama displays!

OTM2166   ISBN: 9781487710361
© On The Mark Press

Make a 'pro' and 'con' chart to show the positive and negative effects of human activities and technologies on the environmental situation you have displayed. Add this to your diorama.

Pros of human and technological involvement	**Cons** of human and technological involvement

Could changing one factor slightly reverse an effect? Explain.

_____

_____

_____

_____

OTM2166   ISBN: 9781487710361
© On The Mark Press

Name:

Making an Impact

Read a newspaper or magazine article that features/ informs about positive or negative examples of human actions or technology on an ecosystem. Summarize the information by completing the form below.

Title of the article: _____

Ecosystem being affected: _____

Describe exactly how the ecosystem is being affected. Is it positive or negative? Use information from the article to support your answer.

_____

_____

_____

_____

Are the changes to the ecosystem a result of human activity, technology, or both?

_____

_____

_____

_____

Discussing the details!

Share your story with a classmate.

OTM2166   ISBN: 9781487710361
© On The Mark Press

Taking Action!

Access the internet to learn more about an organization that works on the principles of sustainable development. Some recognized organizations are:

- Ducks Unlimited
- Nature Conservancy Canada
- Sierra Club
- Earth Island Institute
- National Audubon Society
- Watershed Watch

Choose one of the above organizations or an organization of your own, research their efforts. Answer the following questions:

Organization: _____

What does this organization do? Why? Use information from their website or articles about them to support your answer.

_____

_____

_____

_____

_____

_____

_____

_____

_____

_____

OTM2166 ISBN: 9781487710361
© On The Mark Press

How could **you** take action and contribute to their cause?

_____

_____

_____

_____

Discussing the details!

Share the efforts of this organization with a classmate.

Brainstorm It!

Your next task will be to create a pamphlet to inform the public about the organization and their efforts. Use the space below to outline the information that is important to include on your pamphlet. You will use this to design your pamphlet on Worksheet 9.

What are the threats?	What is the group's action plan to help the cause?	How can others get involved to help?

Create your pamphlet here.

OTM2166   ISBN: 9781487710361

PLANTS – AT THE ROOT OF IT!

LEARNING INTENTION:

Students will learn about the importance of plants in our daily lives and the links between technologies, products, and impacts.

SUCCESS CRITERIA:

- identify the uses of plants in our daily lives
- participate in a food chain game
- describe how pesticide use can affect food chains
- define bioaccumulation and biomagnification, describing its effects on ecosystems
- describe how land use is changing and the effects of this on ecosystems
- research and describe a method used by farmers to increase crop yields

MATERIALS NEEDED:

- a copy of "We Need Plants!" worksheet 1 for each student
- a copy of "A Poisonous Problem!" worksheet 2 and 3 for each student
- a copy of "Trends in our Land Use" worksheet 4 and 5 for each student
- a copy of "Working the Land" worksheet 6 for each student
- a group of three different colored pinnies (for each third of the students to wear)
- many Popsicle sticks (one third of them non-colored, two thirds of them colored)
- access to the internet or library
- clipboards, pencils, pencil crayons, markers, chart paper

PROCEDURE:

*This lesson can be done as one long lesson, or be done in five or six shorter lessons.

1. Give each student worksheet 1 to complete. Upon completion, have a group discussion to share responses to the questions. Through discussion students should realize that we need plants for food, shelter, clothing, air to breathe, medicines. Specific examples should also be discussed.

2. Using a designated area of the school yard or a gymnasium, engage students in a role play of a food chain game.

   a) Begin by spreading many Popsicle sticks on the ground, some non-colored and some colored (about one third colored, and two thirds non-colored).

   b) Assign half of the students to be primary consumers, one quarter of the students to be secondary consumers, and one quarter of the students to be tertiary consumers. Distinguish by using different colored pinnies. Give each of the primary consumers a paper bag to gather food.

   c) Send primary consumers out first, (for about 30 seconds) to gather food (Popsicle sticks).

   d) Send the secondary consumers out to capture primary consumers (for about a minute). Once a primary consumer is captured, the secondary consumer takes its bag and the primary consumer moves off to the sidelines.

   e) Finally, send in the tertiary consumers. They will try to capture any remaining primary consumers and the secondary consumers (for about a minute). Once a consumer is captured, the tertiary consumer takes its bag and the captured consumer moves off to the sidelines.

   f) At the end of the game, ask students still in the game holding bags, to count how many non-colored and colored sticks they have. Tell them that a pesticide was sprayed on the food supply and the colored sticks are poisonous. The more colored sticks they have, the larger the amount of poison they ingested. This could lead to sickness or death in real species.

Discuss with students the following concept:

*Pesticides are herbicides used to control unwanted weeds, and they are insecticides used to control unwanted insects. They are poisonous. Ingesting them makes living things sick. The more they eat, the more the poisonous chemicals accumulate in their cells. This is called **bioaccumulation**. If other living things higher up the food chain, eat animals that have ingested pesticides, the poison accumulates in their cells. This is called **biomagnification**.

3. To enhance students' learning of bioaccumulation, show them 'Bioaccumulation/Biomagnification Movie', episode can be accessed at www.youtube.com.

4. Engage students in a discussion about why bioaccumulation and biomagnification is so dangerous to ecosystems in our world. What can happen if we continue to use pesticides on our plants (agriculture, forestry, home gardens)?

5. Give students worksheets 2 and 3 to complete.

6. Accessing the internet, students will watch 'The Agenda', "The Costs of Saving Farmland", episode can be accessed at www.tvo.org/programs/the-agenda-with-steve-paikin. Click on the search icon, then type in the title of the episode. A follow-up class discussion is optional.

7. Give students worksheets 4 and 5 to complete.

8. Give students worksheet 6, and a pencil and clipboard. With access to the internet, students will research a practice that farmers use today to maximize their crop yield. A follow up option is to have students share their learning with a classmate.

DIFFERENTIATION:

Slower learners may benefit by:

- working with a strong peer to complete worksheet 1

- working in a small group with teacher direction to complete worksheets 4 and 5, answers could be recorded onto one chart paper, which could later be reviewed with the larger group

- working with a strong peer to research the information needed to complete worksheet 6, completion of the worksheet could be done individually or together

For enrichment, faster learners could create a three dimensional model to showcase how a farming technology works to maximize land usage.

OTM2166   ISBN: 9781487710361
© On The Mark Press

Name:

We Need Plants!

There are many interactions in different ecosystems going on in our world, and plants are often at the root of it!

Why do we need plants? What do they give us?

Plants give us _____.

For example, _____

_____

_____

_____

Plants give us _____.

For example, _____

_____

_____

_____

Plants give us _____.

For example, _____

_____

_____

_____

Plants give us _____.

For example, _____

_____

_____

_____

Plants give us _____.

For example, _____

_____

_____

_____

Plants give us _____.

For example, _____

_____

_____

_____

OTM2166   ISBN: 9781487710361
© On The Mark Press

A Poisonous Problem!

Define bioaccumulation.

_____

_____

_____

Draw a diagram that illustrates bioaccumulation.

Describe how bioaccumulation occurred in the living things during the food chain game that you played.

_____

_____

_____

OTM2166   ISBN: 9781487710361
© On The Mark Press

Name:

Define biomagnification.

_____

_____

_____

_____

_____

Draw a diagram that illustrates biomagnification.

Describe how biomagnification occurred in the living things during the food chain game that you played.

_____

_____

_____

_____

Trends in Our Land Use

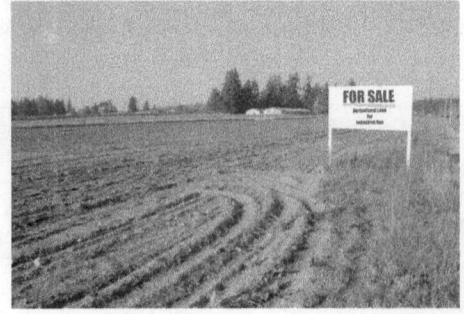

What is the difference between
farmland and cropland?

_____

_____

_____

_____

_____

According to economist and ecologists, is land use
changing today? Explain.

_____

_____

_____

_____

There is a decline in farmland but an increase in
cropland. Where is the increase in cropland gained
from? Explain this concept.

_____

_____

_____

_____

OTM2166   ISBN: 9781487710361
© On The Mark Press

Name:

What technologies are giving farmers the opportunities to have cropland, instead of pastureland?

_____

_____

_____

_____

Draw a diagram to explain how one of these technologies works.

In your opinion, how could the trend in land use from natural environments to managed environments affect ecosystems?

_____

_____

_____

_____

Name:

Working the Land

The trend in land use is moving from natural environments such as grasslands and forests, to managed environments, such as gardens and greenhouses. Your task now is to research a method that is used by farmers to **increase yields** in crop production.

Research It!

What is the process called? Describe how it works.

_____

_____

_____

What are the benefits of this process to society?

_____

_____

_____

_____

Are there costs of this process to the environment? Explain.

_____

_____

_____

_____

OTM2166   ISBN: 9781487710361

SOILS AND PLANT GROWTH

LEARNING INTENTION:

Students will learn about the components in soils, their compaction and moisture content; they will learn about the main parts of a plant, seed germination, and how plants grow to maturity.

SUCCESS CRITERIA:

- make and record observations of the components of different soil types
- investigate the compaction and moisture content of different soils
- make a conclusion about the components and characteristics of soils
- determine the materials or technologies that can enhance soils
- describe the function of the roots, stem, leaves, and flowers on a plant
- examine and identify plant roots as either tap or fibrous
- observe and illustrate the germination process of a bean seed
- track and graph the growth of a bean plant
- describe life cycle of seed plants
- research and describe how seeds germinate and grow to maturity in the natural world
- research and describe techniques used in the farming industry to germinate and grow plants in mass quantities
- identify working parts of a microscope and view prepared slides under a microscope

MATERIALS NEEDED:

- a copy of "Types of Soils" worksheet 1 for each student
- a copy of "Exploring the Soils" worksheet 2, 3, 4, 5 for each student
- a copy of "The Sum of All Parts" worksheet 6 and 7 for each student
- a copy of "What Type of Root Is It?" worksheet 8 for each student

- a copy of "Grow Your Own Bean Plant" worksheet 9, 10, 11, and 12 for each student
- a copy of "Nature vs. Nurture" worksheet 13 for each student
- a copy of "Under the Microscope" worksheet 14 for each student
- soils such as sand, clay, silt, loam, topsoil,
- magnifying glasses, newspaper, toothpicks, measuring cups
- a jug of water, a small gardening shovel, 4 or 5 potting planters (per group of students)
- four different plants with root systems exposed (a set for each group of students)
- bean seeds, plastic Ziploc bags, planting soil, a couple of water cans, a few measuring tapes, a few spray bottles, paper towels
- a potting planter to grow bean plant (one for each student)
- access to water
- access to the internet or local library
- microscopes and prepared specimen slides
- clipboards, rulers, chart paper, markers, pencils, pencil crayons

PROCEDURE:

*This lesson can be done as one long lesson, or done in seven or eight shorter lessons.

1. Using worksheet 1, do a shared reading activity with the students. Along with the content, discussion of some vocabulary words would be beneficial for students' understanding.

   Some interesting vocabulary words to focus on are:

humus	moist	grains
minerals	nutrients	particles

2. Create soil stations in the classroom where students can explore the components of different soil types. Divide students into groups and assign them to a soil station. Give

OTM2166   ISBN: 9781487710361
© On The Mark Press

them worksheet 2. Read through the question, materials needed, and what to do sections with them to ensure their understanding. Give each student worksheet 3 and the materials to begin the exploration using the toothpicks and magnifying glasses. They will also explore each soil type with their sense of touch. Encourage students to share and compare with their group what they have discovered about the different soil samples.

3. Continuing to explore soil types, give students worksheet 4. Give each group a measuring cup, a jug of water, a small gardening shovel or spoon, and a potting planter for each soil type. Students will experiment with soil compaction and moisture content of each soil. Upon completion, students will record their findings about each soil type. *This experiment will span over one or two days.

4. Give students worksheet 5. They will make some concluding statements about what they have learned about soils. With access the internet or local library, they will research what materials or technologies enhance our soils. As a discussion item, pose this question: What soil would be best for growing plants in?

5. Divide students into pairs. Give them worksheet 6. They will engage in a "think-pair-share" activity to discuss and then record answers to the questions on the worksheet. A follow up option is to come back together as a large group to share responses.

6. Give students worksheet 7. Read through with the students about the main parts of a plant and each of their purposes. Students are asked to describe the structure of two different local plants, such as differences in their leaf size or texture, stem size, flower/non-flowering type, root system (if known).

7. Divide students into small groups, and give each of them worksheet 8, four different plants with roots exposed, and magnifying glasses. Students will name and draw each plant, and determine if each plant has a tap or fibrous root system.

8. Give students worksheets 9, 10, 11, and 12, and the materials to germinate their own bean seeds. Students will record observations of their beans' germination. Upon germination, the students will plant their bean seeds and monitor their plant's growth through graphing observations and forming conclusions. They will also create a diagram of its life cycle. *This investigation will span over a number of days.

9. Give students worksheet 13. With access to the internet, students will research and report on how germination and plant growth occurs and spreads in nature, and the techniques used in the farming industry to germinate seeds and grow plants in mass quantities.

10. Using a microscope as a model, instruct students on how to operate its parts appropriately in order to view slides of specimens. Give students worksheet 14, and access to a microscope. They will choose two prepared slides to view under the microscopes. They will record their observations by providing written descriptions of the specimens and diagrams of them. Encourage students to share their magnified specimens with a classmate by viewing them under the microscope.

DIFFERENTIATION:

Slower learners may benefit by:

- working together as a small group with teacher support to germinate and track the growth of a bean plant using worksheets 9, 10, 11, and 12. This would produce one set of data to graph on the line graph. The life cycle diagram could be done together on chart paper.

- pairing up with a strong student to complete worksheet 13

For enrichment, faster learners could research about beneficial and harmful roles of micro-organisms both in natural ecosystems and in humans.

OTM2166 ISBN: 9781487710361
© On The Mark Press

Name:

Types of Soils

Soil is made up of tiny rocks, water, air, and humus. When rocks experience a change in temperature, they crack into smaller pieces. When wind and water hit against rocks over time, they crack into smaller pieces. Tiny rocks and minerals become part of soil.

What is humus?

Humus is made when rotted plants and animal matter break down over time into smaller pieces. The humus mixes with tiny rocks, water, and air to make soil. Humus provides nutrients and keeps the soil moist.

There are different types of soils, such as sand, clay, silt, loam, and topsoil. Let's read more about these!

Sand is made of tiny rocks and mineral pieces like quartz. The particles in sandy soil are bigger and loose. Sand does not hold water or nutrients very well.

Clay has fine rock particles in it. The rock grains in clay are small and close together. Clay holds water well, but there is not much air in it for plants to grow well.

Silt is made of sandy soil and clay. It holds water well, but is easily blown away by wind and washed away by water.

Loam is made of sandy soil, clay, and silt. Loam can stay moist because water and air can flow through its particles.

Topsoil is made up sandy soil, clay, and humus. It holds water very well. The humus in it makes it rich in nutrients.

Exploring the Soils

Question: How would you describe the different types of soils?

Materials Needed:

- newspaper
- a magnifying glass
- a measuring cup
- access to the internet
- different types of soils such as sand, clay, silt, loam, and topsoil

- toothpicks
- 5 potting planters
- access to water
- small gardening shovel or spoon

What To Do:

1. Using the magnifying glass, some toothpicks, and your sense of sight, describe how each of the soils look. Record your answers in the chart on worksheet 3.

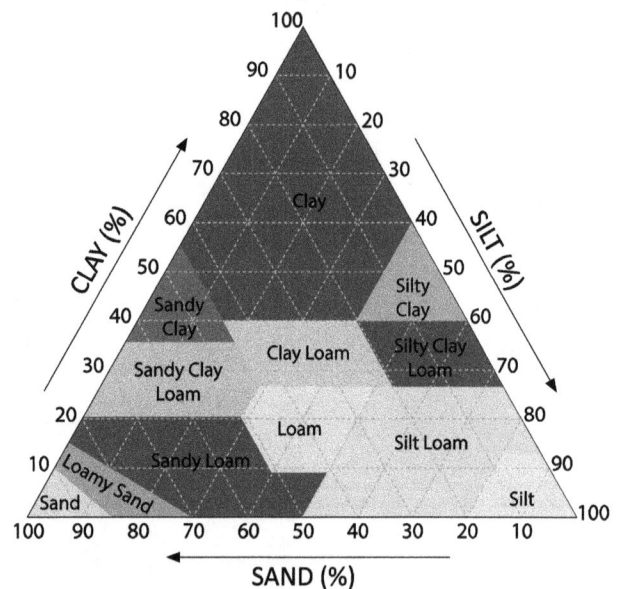

2. Use your sense of touch to describe how each of the soils feel. Record your answers in the chart on worksheet 3.

3. Measure out the same amount of each soil and place each into a potting planter. Compact each soil type down as much as possible. What do you notice? Record your observations on worksheet 4.

4. Measure out the same amount of water and add it to each of the soils in the potting planters. Leave it for a day or two. What do you notice? Record your observations on worksheet 4.

5. Make some conclusions about soil. Record them on worksheet 5.

6. Make some connections about soils and our environment. Record them on worksheet 5.

84

OTM2166   ISBN: 9781487710361
© On The Mark Press

Let's Explore!

Describe how each of the soils look and feel.

Soils	Describe what it looks like and feels like
Sand	
Clay	
Silt	
Loam	
Topsoil	

OTM2166   ISBN: 9781487710361
© On The Mark Press

Measure out the same amount of each soil and place each into a potting planter. Compact each soil type down as much as possible. What do you notice?

After compacting down each of the soils, I noticed...

_____

_____

_____

_____

_____

_____

Measure out the same amount of water and add to each potting planter. Leave it for one or two days. What do you notice about the moisture content?

After one or two days, I noticed...

_____

_____

_____

_____

_____

_____

OTM2166   ISBN: 9781487710361
© On The Mark Press

Name:

Let's Conclude

What have you discovered about soil? What conclusions can you make?

_____

_____

_____

_____

Let's Connect It

Why do **you** think soil is so important to our environment?

_____

_____

_____

_____

Access the internet to research what materials or technologies can enhance our soils. Record your findings.

_____

_____

_____

_____

OTM2166 ISBN: 978148771036l
© On The Mark Press

Name:

The Sum of All Parts

Think Pair Share

With a partner, do some thinking and
sharing of ideas about the questions below.
Record your ideas.

"What does a plant use its roots for?"

"What does a plant use its stem for?"

"What does a plant use its leaves for?"

"What does a plant use its flowers for?"

OTM2166 ISBN: 9781487710361
© On The Mark Press

Fast Facts!

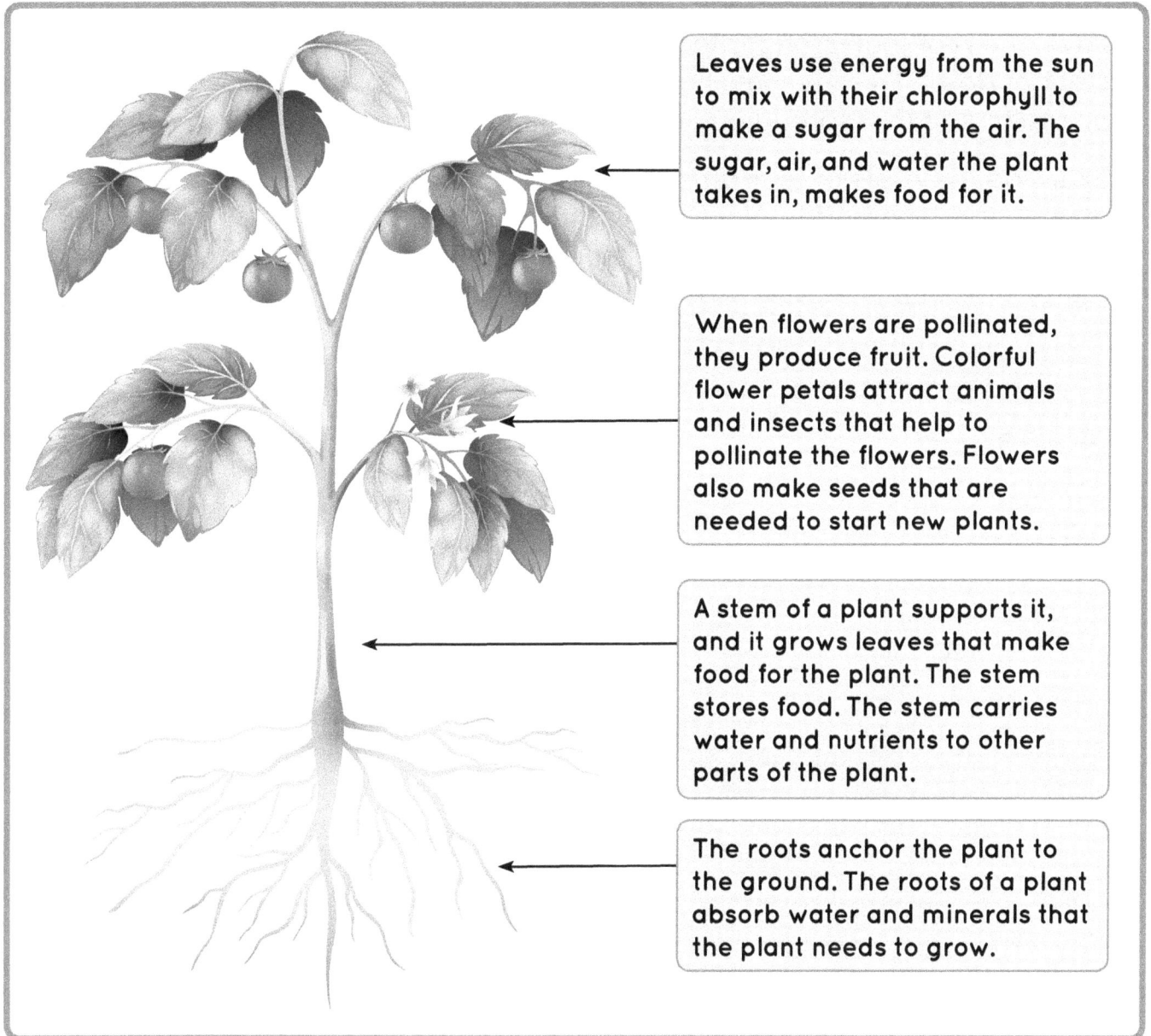

Leaves use energy from the sun to mix with their chlorophyll to make a sugar from the air. The sugar, air, and water the plant takes in, makes food for it.

When flowers are pollinated, they produce fruit. Colorful flower petals attract animals and insects that help to pollinate the flowers. Flowers also make seeds that are needed to start new plants.

A stem of a plant supports it, and it grows leaves that make food for the plant. The stem stores food. The stem carries water and nutrients to other parts of the plant.

The roots anchor the plant to the ground. The roots of a plant absorb water and minerals that the plant needs to grow.

Choose two different plants that are common in your area. Describe how they are different in structure.

_____

_____

_____

_____

What Type of Root Is It?

Roots are either tap or fibrous. **Tap roots** are roots that have one root that is longer than the rest, and it grows straight down. **Fibrous roots** have many strands of roots of similar size that spread out in all directions.

Examine the roots of four different plants. During your examination, name the plant, draw the plant's root system, and determine its root system type, (circle tap or fibrous).

Name: _____

tap   fibrous

Name: _____

tap   fibrous

Name: _____

tap   fibrous

Name: _____

tap   fibrous

OTM2166   ISBN: 9781487710361
© On The Mark Press

Name:

Grow Your Own Bean Plant!

Have you ever wondered how a seed grows into a plant? In this experiment you will watch a bean seed grow into a plant. Let's grow!

Materials Needed:

- about 5 bean seeds
- a plastic Ziploc bag
- spray bottle
- a paper towel
- a planter

- water
- watering can
- a measuring tape
- a small gardening shovel
- soil

What To Do:

1. Using the spray bottle of water, mist the paper towel on both sides.

2. Put the bean seeds on the paper towel and fold it over a couple of times so that the bean seeds are covered.

3. Place the paper towel with the seeds in it, into the plastic Ziploc bag, and seal it. Leave it for 3 days.

4. After 3 days, check on your seeds. Has there been any change? On worksheet 10, record your observations.

5. Use the spray bottle to mist the paper towel, then place it back into the Ziploc bag. Leave it for another 3 days.

6. Repeat steps 4 and 5 if your seeds have not germinated yet.

7. After germination has begun, plant the seeds in a planter of soil. Put it in a sunny spot. Water your plant every 3 days, and measure its growth.

8. Graph your observations of your plant's growth and make conclusions about its growth on worksheet 11.

9. Create a diagram of a seed's life cycle on worksheet 12.

OTM2166 ISBN: 9781487710361
© On The Mark Press

Name:

Let's Observe

Illustration of the seeds after 3 days in the moist towel:	Illustration of the seeds after 6 days in the moist towel:	Illustration of the seeds after 9 days in the moist towel:

Tell about your seeds' germination process:

_____

_____

_____

_____

_____

_____

_____

OTM2166   ISBN: 9781487710361
© On The Mark Press

Name:

Graph the Growth!

93

Create a line graph to display the growth of your bean plant.

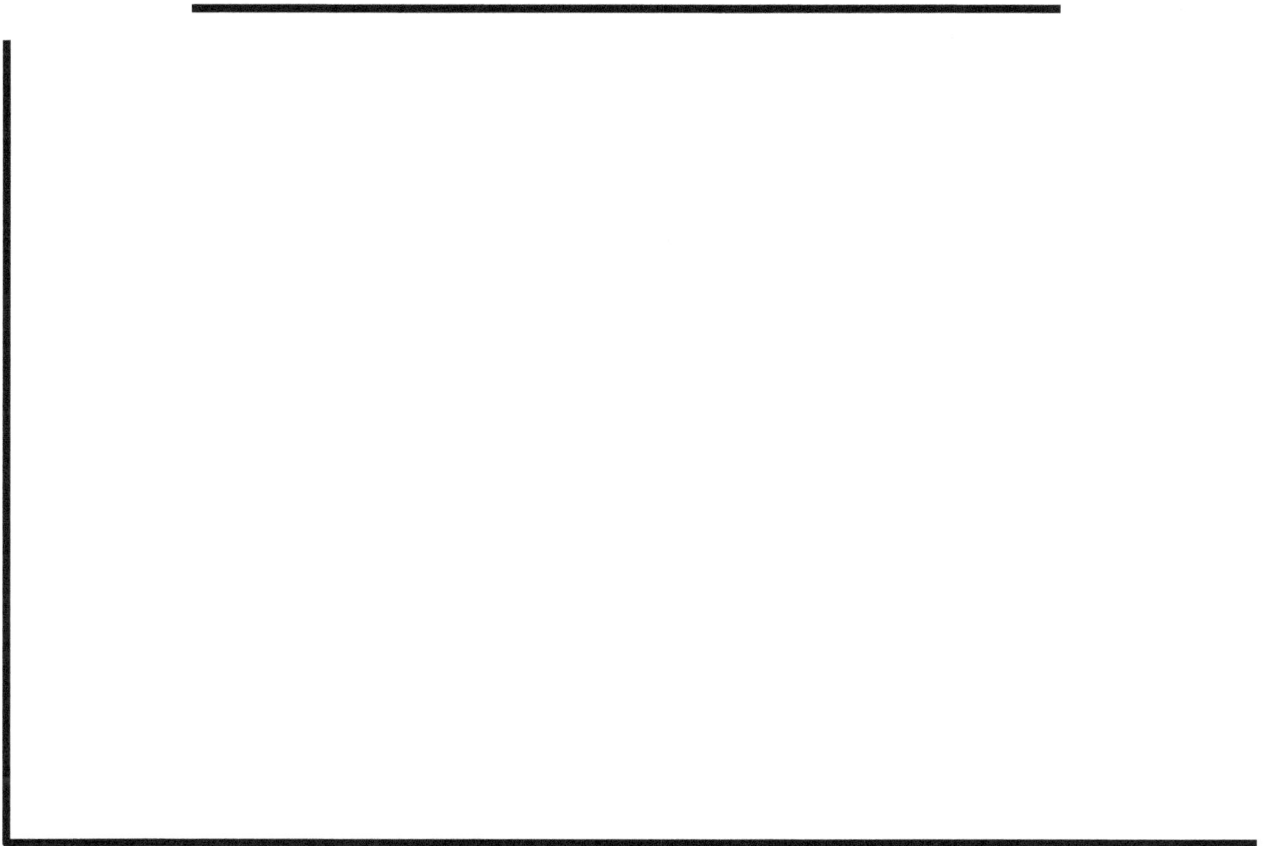

Let's Conclude

What conclusions can you make about your plant's growth?

_____

_____

_____

_____

OTM2166   ISBN: 9781487710361
© On The Mark Press

Name:

Let's Connect It!

Create a diagram to explain the life cycle stages of a bean plant.

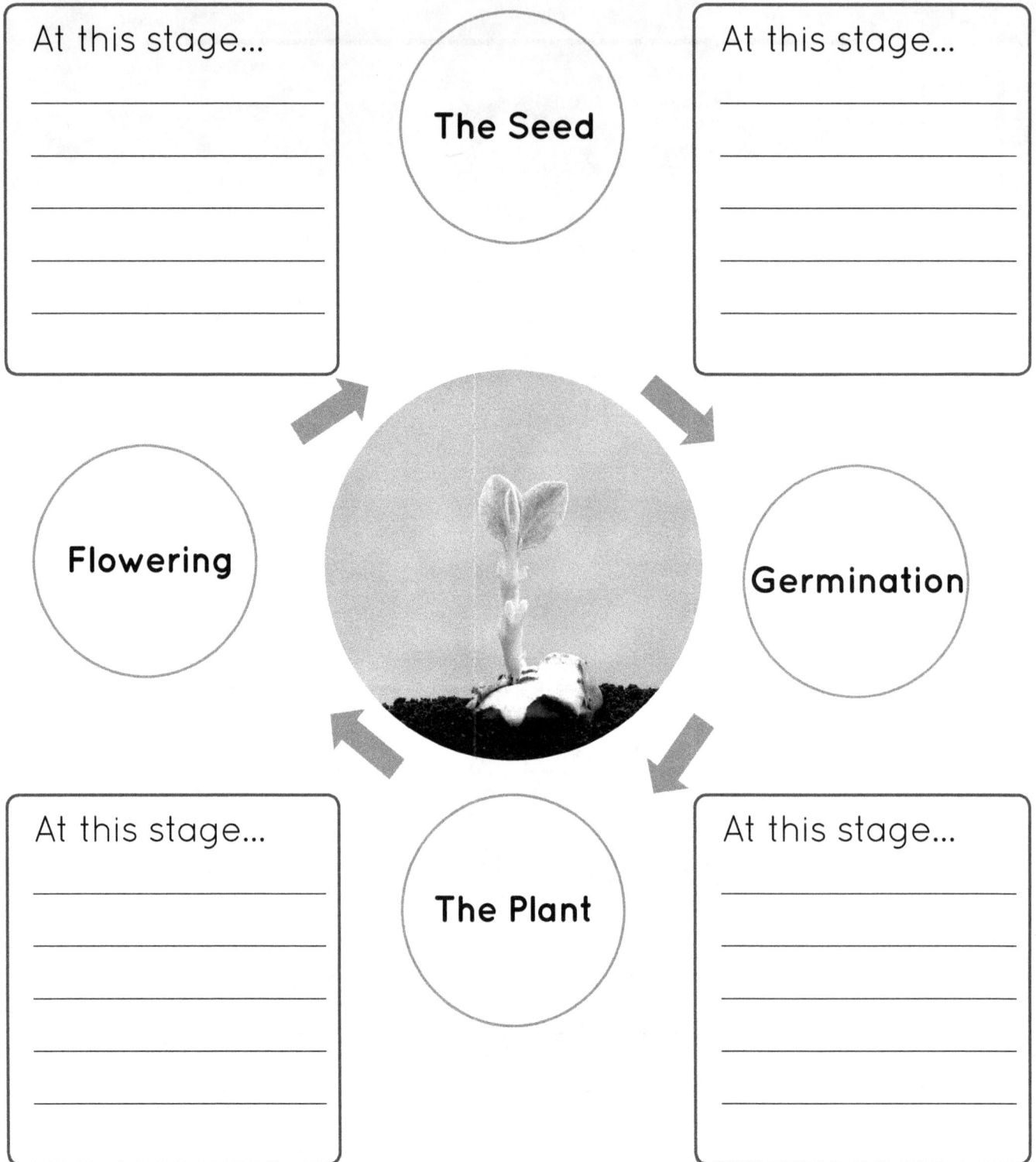

At this stage...

_____

_____

_____

_____

The Seed

At this stage...

_____

_____

_____

_____

Flowering

Germination

At this stage...

_____

_____

_____

_____

The Plant

At this stage...

_____

_____

_____

_____

OTM2166   ISBN: 9781487710361
© On The Mark Press

Name:

Nature vs. Nurture

Let's Research It!

You have learned how to germinate a seed, to ready it for planting. Have you wondered how this process occurs in nature? Conduct some research to find out about this natural process.

How do seeds germinate and continue to grow and spread in nature?

_____

_____

_____

_____

_____

_____

You germinated one seed, to ready it for planting. Have you wondered how farmers grow an abundance of plants from seeds and the techniques they use to produce in mass quantities? Conduct some research to find out about this nurturing process.

What techniques are used in the farming industry to germinate seeds and grow plants in mass quantities?

_____

_____

_____

_____

_____

OTM2166 ISBN: 9781487710361
© On The Mark Press

Name:

Under the Microscope

Slide 2

Specimen: _____

Magnification: _____

Diagram:

Detailed description:

_____
_____
_____
_____

Slide 1

Specimen: _____

Magnification: _____

Diagram:

Detailed description:

_____
_____
_____
_____

OTM2166   ISBN: 9781487710361

www.ingramcontent.com/pod-product-compliance
Lightning Source LLC
Chambersburg PA
CBHW082108210326
41599CB00033B/6626